Race and Racism

Biology, Ethics, Politics

Michel Tibayrenc
MIVEGEC, Univ. Montpellier, IRD, CNRS
Montpellier, France

CRC Press
Taylor & Francis Group
Boca Raton London New York

CRC Press is an imprint of the
Taylor & Francis Group, an **informa** business
A SCIENCE PUBLISHERS BOOK

Cover by Myrtille Tibayrenc; www.myrtilletibayrenc.com (inspired from a photography by Dirk Skiba).

First edition published 2027
by CRC Press
2385 NW Executive Center Drive, Suite 320, Boca Raton FL 33431

and by CRC Press
4 Park Square, Milton Park, Abingdon, Oxon, OX14 4RN

CRC Press is an imprint of Taylor & Francis Group, LLC

Library of Congress Cataloging-in-Publication Data (applied for)

ISBN: 978-1-041-08273-6 (hbk)
ISBN: 978-1-041-08890-5 (pbk)
ISBN: 978-1-003-64739-3 (ebk)

DOI: 10.1201/9781003647393

Typeset in Times New Roman
by Prime Publishing Services

Contents

Methodological Note

Words marked with an asterisk* refer to the glossary of specialized terms and abbreviations. They are only marked as such the first time they appear in the text.

This book is based on an in-depth and thoroughly updated analysis of the literature relating to the biological diversity of human geographical populations and the ideologies connected to it. It gives prominence to "classic" scientific publications, meaning those found in international, peer-reviewed scientific journals. The professional audience (researchers, teachers, students) will therefore find useful and highly current information on this debate. However, more informal media constitute, in my view, a very valuable complementary source of information. The reader will therefore find numerous references to more "general public" publications, particularly excellent articles gleaned from "leading" media, which host talented scientific writers. I have not disdained, quite the contrary, the Wikipedia source, as I am not very receptive to Elon Musk's criticisms against this medium. Undeniably, Wikipedia has, if not an editorial line, at least a general progressive tendency, which probably reflects the general profile of its contributors. Nonetheless, this medium generally appears to me to be well-documented, with the onus on the user to sort through the information. Wikipedia, and social networks in general, are like Aesop's tongue: they combine the best and the worst. It is a question of how they are used.

A number of references come from French sources. But the English-speaking reader knows that Google Translate, Google Gemini, and ChatGPT work miracles.

Finally, I have not refrained from analyzing works that have been highly controversial, to the extent that the information they contained appeared reliable to me.

My objective has been to make the book as accessible as possible to the lay public. Specialized jargon has been avoided as much as possible. A glossary of specialized terms has been added to aid in reading. For the non-specialist audience who might be put off by overly obscure scientific developments, the important points of each part of the book are summarized as they appear in italicized boxes.

This Book is Definitely Based on a Scientific Approach, With an Emphasis on the *Biological* Characterization of Geographical Populations

"Science" is taken here with a broad meaning. Human sciences are unambiguously included. The criteria for defining what constitutes a scientific approach are the subject of never-ending controversies. I adhere here to a simple definition: a process of objective knowledge based on rational reasoning and reliable data according to the current state of knowledge in the discipline concerned, under the constant criticism of peers, with results published in high-quality specialized media. This naturally applies to the "classic" scientific documentation in this book. For more informal sources of information, greater flexibility and distance are welcome. However, even for this literature, I have been careful in my selection, and have certainly not retained certain very low-quality "information" that is rampant on many social networks.

As with the previous work "What Makes us Humans" (Tibayrenc and Ayala, 2020), I consider it crucial to constantly sort between what is: (i) robust, satisfactorily supported; (ii) simply probable; (iii) what remains in the realm of the speculative; (iv) what we do not know (yet); (v) and what is obviously false. The reader will note that I exclude "sure" and "certain" from this inventory. These terms should be regarded with suspicion in science, a field in which only working hypotheses, provisional by nature, should be retained. Of course, I have little doubt that the Earth is not flat. However, this is an extreme case, and most scientific knowledge that the public considers "sure" are only working hypotheses that are satisfactorily supported until proven otherwise.

The Most Reliable and Robust Approach for Characterizing Geographical Populations: Population Genetics, Population Genomics*, Evolutionary Genetics*, Phylogenetics*.*

In this perspective, the "backbone" of the book is what constitutes my personal area of expertise: apart from my specific training in tropical medicine, on which I relied for the "diseases" section, my specialization as a scientific researcher is in population genetics, population genomics, evolutionary genetics, and phylogenetics. More specifically, I have focused on the identification of "relevant units of analysis" (Tibayrenc and Ayala, 2017a) in microbial populations.

It turns out that these disciplines, which I have specialized in, allow for the very reliable elucidation of migrations, genealogies, and degrees of kinship among human populations. With this approach, one is very close to the direct observation of raw data. In contrast, what relates, for example, to the heritability* of psychic traits or brain function remains much more speculative.

I confess that I am not primarily a specialist in human biology. However, this was also the case for the famous evolutionists Theodosius Dobzhansky and Ernst Mayr, who were a *Drosophila* (fruit fly) specialist and an ornithologist, respectively. This did not prevent these two illustrious scholars from writing very relevant reflections on human geographic populations.

In the pioneering era of population genetics, the 70s and 80s, the scarcity of reliable genetic markers and their lack of resolution necessitated calling upon a mass of statistical and mathematical analyses whose reliability was hampered by numerous, often speculative, working hypotheses. The impressive technical progress of the last twenty years, particularly in the fields of genomics*, big data computing, and medical imaging, has allowed for the advent of a *"neo-Pasteurian era"* (from the French microbiologist Louis Pasteur, 1822–1895), i.e., where the direct observation of a mass of raw data has become prevalent again. This has considerably increased the reliability and informative power of population genetics and genomics, as well as phylogenetics. Our knowledge of human biological variability has been multiplied, which significantly clarifies the subject of the present book.

The most important part of this book therefore focuses on what characterizes geographical populations *on a biological level*. This is a subject about which many false ideas circulate, and which requires an exhaustive clarification.

"Scientific Truth"?

There is no "scientific truth"; There are only hypotheses.

This crucial point has already been mentioned above. In my eyes, and this appears to me to be quite important, there is no "scientific truth". As soon as one moves, even slightly, away from raw observation, there can only be *working hypotheses and models*, that are provisional by nature. The French neurobiologist Hervé Chneiweiss (2019) states that "a scientific fact is true or wrong". This position seems to me excessive, even simplistic. I would say rather that a scientific hypothesis is more or less solidly supported.

The general public, including seasoned journalists, find it difficult to accept this uncomfortable idea of "scientific uncertainty". Yet, it is the guarantee of rigor and flexibility of the scientific approach. "Scientific truths" are not set in stone. This remark is important for the theme of this book.

"Only the truth is revolutionary," wrote the Italian Marxist theorist Antonio Gramsci. This is a pertinent thought, which many "revolutionaries" could make better use of. However, we, scientists, will not propose, or should not propose, "truths"; only *models* that hold up less badly than others.

Popper's deductive approach: only a falsifiable hypothesis is scientific.

Popperism

Popper's criterion of falsifiability* (Popper, 1973), or the deductive approach, although still relevant, appears to me too strict to be operational in many cases. This criterion considers as scientific only hypotheses that are *falsifiable*.

Inductive approach: presumptions are accumulated till they are considered sufficient.

Regarding medical research and medicine particularly, hypotheses are often formulated through an *inductive process* with no other ambition than to amass premises and presumptions that are considered sufficiently convincing. One might consider that such an inductive approach is philosophically unsatisfactory. However, it is efficient, which is what matters. In the fields of medicine and biomedical research, because decision-making is crucial, we cannot afford to limit ourselves to a strict Popperian approach, which one of my very facetious and provocative colleagues qualified as "Popperian obscurantism" (Ninio, 1979), and which a professor of the great ethologist Konrad Lorenz (I forgot the source) considered "the height of the most narrow-minded dogmatism". The Popperian criterion can also be faulty, as a non-false hypothesis may be wrongly falsified based on erroneous bases, which are always possible. Hypotheses based on such an inductive approach will therefore be frequently cited in this work.

Down with Propaganda and "Fake News"

Fake news": a very difficult concept to define.

The subject addressed by this book is ideologically and politically sensitive. Stoczkowski (2007) noted that the question of racism is one of those that particularly encourage the display of good intentions. Moreover, there is no doubt that ill intentions are by no means left out. Faced with this extremely polarized debate, the temptation is great for propagandists, whether well- or ill-intentioned, to flood the public with biased information. Social networks (Facebook, YouTube, X, TikTok, Instagram, etc.) are a major source of unfiltered information whose reliability is questionable. These networks provide a free and unlimited drug to the rampant followers of conspiracy theories. I am by no means saying that all information from these social networks should be dismissed. I simply think it must be used with extreme caution. On these networks, one finds multiple "proofs" that the Earth is flat, and that Americans never went to the moon.

"Leading" media establish sections intended to detect unreliable information ("The decoders" of the French daily newspaper *Le Monde*, for example). French President Emmanuel Macron even proposed to enact a new law to condemn the disseminators of "fake news". The academic world has raised the alarm about the necessity of better countering the phenomenon of "fake news" (Lazer et al., 2018). However, the definition of "fake news" given by the authors of this article is perplexing. According to them, any source of information outside the major media is suspected of disseminating "fake news," while these major media such as the *Wall Street Journal* and the *Washington Post* would be decreed reliable a priori and should be entrusted with the sacred mission of hunting down "fake news". Lazer et al. (2018) ask: "how can we create a new ecosystem and a new culture that value and promote the truth?" They simply forget to define what "truth" is (see my remark above on "scientific truth").

Few facts can be considered "true", apart from verifiable numerical data.

Lazer et al. (2018) misunderstand the essential fact that beyond raw data, if possible numerical (the population of a country on a given date, or Donald Trump is 6.23ft tall, for example), which are entirely verifiable, the world of information is an immense gray area in which "facts" are subject to interpretation: "Truth on this side of the Pyrenees, error beyond" (the French philosopher Blaise Pascal). This gray area is a windfall for the manufacturers of "fake news", who shamelessly exploit its ambiguities to inundate us with obscured half-truths and sophisms. It is therefore extremely difficult and crucial to define "fake news," especially because many examples (whose obvious goal is to mislead the reader—the very definition of sophistry) are cleverly disguised.

Counter-truths and flat lies are quite easy to refute. Skilled sophisms are not. Furthermore, it would be supremely naive to believe that the major media and even classic scientific literature are ideally neutral and exempt from any false information whatsoever. Could one not suspect that the sections specialized in hunting "fake news" advocated by Lazer et al. (2018) are an attempt by "mainstream" media to reserve an exclusivity on "fake news"?

Therefore, it is futile to state a "scientific truth." On the other hand, it is possible to unearth "scientific counter-truths," statements that prove to be perfectly false.

Involuntary, but (Even More?) Damaging "Fake News"

I advocate for the rehabilitation of a certain modest and cautious science, sticking as closely as possible to the original texts cited in extenso, to raw observations, and to a simple description of the world. As few "models" as possible, as many raw facts as possible. The assertoric (simple description of facts) must precede the apodictic (establishment of necessary relationships) (Fabre, 1999). As said above, the progress of modern tools in genomics, big data computing, and medical imaging allows for the advent of a *"neo-Pasteurian era"*, in which raw observation reigns supreme again.

Prioritize Raw Information over Speculative Data. Combat the "Myside Bias".

The lay public does not differentiate between concrete facts that are robustly supported and entirely speculative science, which is currently in vogue ("it's all science"). Today, this speculative science is literally riddled with involuntary (not always) "fake news," which are quite often refuted by raw facts. For example, it was long considered, even in textbooks and reference works, that the number of human genes was greater than several hundred thousand, without the slightest proof. Progress in genomics has shown that this number is much lower, around 20,000. This kind of gratuitous statement, typical of involuntary "fake news", repeated indefinitely, is quite damaging to the credibility of science and is abundantly exploited by its detractors. Involuntary "fake news" is aggravated by the "myside bias", which pushes us to select, in the facts, those that reinforce our convictions and to deny or conceal the others (Orlando, 2021). This bias is all the more dangerous because the "cognitive elites," intelligent and educated people, or those presumed to be, are by no means spared.

Anti-"Fake News" Toolkit: Return To Original Writings and Data

Whether on social networks, in the general public media, or even in scientific literature, propagandists constantly use the same tricks:

(1) Omitting to cite original sources or to return to them, and preferably using second or third-hand citations. For example, it is broadly accepted that the Soviet scientist Trofim Lysenko was an obscurantist, who undermined Soviet

science for a very long time. This is very likely true. Now, who has read Lysenko extensively in the original Russian text? Very few people, certainly. Not me, in any case.

(2) Using the "strawman strategy" by attacking a given theory, not for what it truly is, but for erroneous and distorted statements of that theory. This point will be extensively developed in the discussion on scientific antiracism* (Part I, Chapter 1). An effective remedy against this strategy consists of broadly using original sources and complete quotations. Words fly, writings remain, fortunately for science.

(3) Isolating a particular case to make it seem generalizable, while avoiding all comparisons.

Avoid Confusing Morality, Ethics, and Politics on One Hand, With Scientific Knowledge on the Other ("Reciprocal Sanctuarization"; Tibayrenc, 2019)

This is a major theme of the present book. I will return to it in detail in the general conclusion.

Introduction

The Human Rainbow in the Light of Genetics

This book is the culmination of a long intellectual and scientific journey that began in the 1970s. At that time, I was a young researcher already specializing in the genetics and evolution of pathogens and had just been recruited by the IRD (Institute for Research for Development).

In keeping with the mission of this French governmental research agency, I was posted overseas and abroad for many years, accompanied by my wife and children. My life outside continental France—which began with a year as a medical doctor for SONATRACH (the national gas and oil company) in Algeria—gave me a very concrete understanding of the human rainbow, or the diversity of our species*.

This experience includes one year in Algeria, one year in French Guiana, seven years in Bolivia, five in the United States, and three in Thailand, not to mention multiple shorter trips to sub-Saharan Africa and various parts of Latin America.

My dual journey—extensive contact with highly diverse populations on the one hand, and my scientific specialization on the other—allowed me to deeply nourish my reflection on the human rainbow. The various countries of assignment provided me with very diverse experiences, disparities only heightened by the fact that you do not see the world in the same way at ages of 25 and 78.

In a way, I have had experiences similar to those of the "father of American cultural anthropology," Franz Boas (1858–1942), who forged his convictions following numerous study trips to diverse peoples. Like him, I humbly adopted the crucial notion of cultural relativism and cured myself of any ethnocentrism.

Below are some reflections and anecdotes on these enriching experiences:

French Guiana: The colorful overseas France

In French Guiana, the dominant group, improperly referred to by themselves as "Creoles" (a Creole in the original sense is a white person born in the French overseas territories, such as Joséphine de Beauharnais, Napoleon's wife, born in the island la Martinique), is made up of the descendants of African slaves, with a whole range of mixed ancestries.

These differing degrees of mixing have given rise to a rich and varied terminology: mulatto (one black, one white parent), quarteroon (one quarter black blood), chabin (a mixed-race person with African features but light skin and blond

Figure 1: The author in the 1980s with Aymara children in the region known as the "Yungas" (subtropical valleys descending toward the Bolivian Amazon).

or reddish hair). This abundance of vocabulary indicates that Creole society has a clear racial awareness, meaning it distinguishes racial origin in the biological sense, without necessarily seeing it as a hierarchy. It is anything but "color-blind."

These Creoles live alongside the "métros" (for "Métropolitains"), meaning French people from mainland France, a fairly large Chinese community, the Hmong—a mountainous Asian ethnic group that fled persecution from the Laotian communist regime—and various communities of Native Americans. Added to all this are Haitian, Brazilian, Surinamese, and Lebanese minorities.

"Plurinational State of Bolivia"

Bolivia is the only country in the Americas where Native Americans are the dominant group, representing more than 70% of the population.

Two main groups of "indigenos" (natives) or more simply "indios," constitute this population: the Aymaras, who are the original inhabitants of the Altiplano (the high plateau situated at 13,123 ft in altitude, straddling Bolivia and Peru), and the Quechuas, who are the descendants of the Inca conquerors. Many Bolivians are trilingual: Spanish, Aymara, and Quechua. Figure 1 shows me in the 1980s in the company of Aymara children, who provided me with invaluable assistance in collecting "vinchucas," the bugs that transmit Chagas disease*.

A third Native American group is made up of the Guarani, in the Santa Cruz region near the border with Paraguay. Various other Amerindian minorities, sometimes limited to a single village, complete the picture. These people, although all Amerindian, identify one another at first glance or by ear.

A significant mixed-race minority, the "Cholos," constitutes a large part of the urban middle and upper class. Unmixed Bolivians of European descent are few in number. Finally, at the time of the Conquista, the Spanish attempted to bring African slaves with them to work in the tin and silver mines of Potosí. However, these Africans tolerated the altitude (Potosí is at 13123 ft) very poorly, not to mention the forced labor and poor living conditions imposed on them.

Their descendants, small in number, took refuge in the mid-altitude valleys with a subtropical climate, the "Yungas" (hot valley in the Quechua language). It is striking to see women of pure African descent elegantly wearing the traditional costume of the Aymara women (the "cholitas"), complete with the famous bowler hat (sombrero hongo) and ample, colorful skirts.

This human rainbow fully justifies Bolivia's official name, "Estado Plurinacional de Bolivia" (Plurinational State of Bolivia). For the political power, the country's different human groups constitute as many "nations," and Bolivia has been able to pride itself on having the first Latin American president of pure Native American ancestry, the Aymara Evo Morales, who, moreover, has a facial structure absolutely typical of his population of origin.

The United States, a still conflictual multiracial society that seeks to improve

There is no need to elaborate on this aspect of the United States, as it is known to all. I have had very diverse experiences in that country: a total of four years in California (University of California Davis and Irvine) and 15 months in the South, in Atlanta (Centers for Disease Control, the famous CDC). The United States is not only a multiracial country but also a highly racialized one, an important aspect for the subject of this book. For many administrative procedures, you are asked (optionally) to fill out a questionnaire regarding your ethnic group, which is infinitely surprising for a French person. So, as far as I was concerned, I dutifully ticked the "Caucasian" box. Moreover, these administrative ethnic categories are the subject of virulent debates, as we will see later. California and the South are very different communities. In California, Latino immigrants, particularly Mexicans—the "Chicanos"—are omnipresent. In the South, particularly in Atlanta, the "African American" community is very large, as is its cultural footprint. Everywhere, both in California and Atlanta, I witnessed the commendable efforts of the administration to promote underrepresented minorities through "affirmative action." The United States is also the birthplace of "woke culture," one of whose striking features is to sniff out racism wherever it may be found, as North American society is supposedly afflicted with "systemic racism," and Americans of European origin are cataloged as inherently racist, even and especially if they claim to be antiracist.

Thailand, a Legally Racialized Society

The Kingdom of Siam hosted me for a three-year stay as the representative of the IRD in Bangkok. I brought back a wife from there, which leads me to spend several months a year in the "Land of Smiles." I have an acceptable command of the language, which is essential for better understanding the local mindset. Thailand, literally, is the "Land of the Thais." The Thai ethnic group is indeed the majority there, making up about 90% of the population. A large minority of Chinese descent, which is very well integrated, is present, especially in the cities. One also finds Indian, Middle Eastern, Burmese, Cambodian, Laotian minorities, ethnic minorities such as the Hmong, a small population of aborigines—probably descendants of the country's first inhabitants—and of course the "farangs," meaning Western Caucasians. Legally, Thailand has the peculiarity of having two categories of nationality: "legal" nationality (san chaat, สัญชาติ), and "ethnic" nationality (chua chaat, เชื้อชาติ). Thus, a Sino-Thai has the legal nationality: "Thai" and the ethnic nationality: "Chinese" (chin; จีน). This is, of course, very surprising for a French person.

The spectacular human diversity of these countries is not limited to the physical aspect but is also reflected in notable biological differences—particularly, in Bolivia, the adaptation to altitude, and in all these countries, susceptibility to diseases (see Part I, Chapter 2).

All these observations "clashed" in my mind as a geneticist with a theory that was strongly spreading at the beginning of my career among biologists, and even more so among social science specialists: *the non-existence of human biological races*. According to this concept, a cornerstone of "scientific antiracism," "races" are purely cultural and sociological creations without any biological foundation. The leaders of this school of thought were the paleontologist Stephen Jay Gould and the geneticist Richard C. Lewontin (recently deceased), both standard-bearers of the left-leaning "Science for the People" movement. In France, the late Albert Jacquard was also a champion of scientific antiracism.

Even at the time, however, there was no consensus on this subject among biologists. Famous evolutionists like Theodosius Dobzhansky and Ernst Mayr, both notorious architects of the "Synthetic Theory of Evolution," defended the existence of the biological reality of human races. The debate has continued to this day. It takes on very different aspects in France compared to the Anglo-Saxon world, and, very remarkably, in the "Western" world compared to Eastern Europe, Russia, and China, which clearly shows that there is no consensus on the existence or non-existence of human races.

As for me, I know a little more than I did at the time, or more precisely, I have a more precise idea of my shortcomings and those of current science. My reflection has continued, enriched by the spectacular developments in evolutionary science and genomics.

Through my professional activity, I have been led to work for more than 40 years on genuine scientific "Chinese puzzles," namely the notions of species, subspecies*, strain*, and clone* among pathogenic agents (viruses*, bacteria*, parasitic protozoa*, yeasts*, and pathogenic fungi*). Compared to these abstruse concepts, the problem of the taxonomic and evolutionary definition of human geographical populations is relatively trivial, at least in the eyes of one who wipes the fog of ideology from their glasses.

This reflection has also been nourished by the fact that, passionate about natural history since childhood, I have racked my brains over the concepts of species and subspecies in fossils, butterflies, and other insects. Finally, as my career path summarizes, I have closely observed and interacted with human geographical populations (whatever name one gives them) for many years. I have a very concrete experience of this. These people welcomed me warmly, I worked with them, I immersed myself in their way of life, and I laboriously learned their languages. This total, physical, visceral immersion is worth far more than a few statistical formulas or a few ethnology books.

Aim of the Present Work

The purpose of this work is to synthesize this long-term reflection and to take stock of our most recent knowledge concerning human genetic diversity and the state of the debate on races and racism. As set out in the "Methodological Note" section, the approach focuses primarily on what constitutes my field of excellence: the genetic and taxonomic characterization of populations. This is what I have done throughout my career regarding populations of pathogenic microorganisms, and to a lesser extent, populations of disease vectors and their hosts, first and foremost, humans.

The goal of this work is in no way to propose a rehabilitation of the biological concept of race. I simply present the different models that are in conflict in the context of this controversy. The reader will form their own opinion. They will see that the controversy is primarily a matter of *definitions* and *semantics*. But semantics holds an exorbitant importance in this debate, which is marked by a tragic past and a highly conflictual present.

In writing this book, I was well aware that by expressing reservations about "scientific antiracism," at least regarding the raw statement that "human races do not exist," there was a risk of giving ammunition to racists and identitarians. However, I decided to dismiss this concern, because in any case, these people will always find something to fuel their vision and do very well without any scientific endorsement. More dangerously, as I have seen while browsing some of their websites, some show skill in pinpointing the weak points of the antiracism doxa that "human races do not exist," just as Jehovah's Witnesses or certain Islamic sites know how to "press where it hurts" on certain overly dogmatic aspects of the theory of evolution.

The danger of the rise of "scientific and medical obscurantism" due to a misunderstood and misguided antiracism appeared to me to be much more pressing than a few identitarian yelps. Faced with obscurantism, scientific rigor, stripped of all ideology as far as possible, remains the best weapon for building effective antiracism. I hope to have observed such rigor in the writing of this book.

My conviction is that all debate is acceptable and welcome, as long as knives and insults remain in the cloakroom. At one conference I was organizing, I had planned to invite scientists who were proponents of "intelligent design" to debate. Alas. To my regret, I was met with an outcry from some conference participants, who openly threatened to boycott the event if I persisted in my intention. The pretext invoked, frequently used by censors, was that by inviting these defenders of intelligent design, I was "offering them a platform," and therefore, giving them legitimacy. In my opinion, it is those who refuse the debate who are in the wrong. If the arguments of intelligent design are scientifically fallacious, there should be no harm in refuting them.

In the context of the present debate, I am open to any dialogue with scientists who do not share my point of view, or even with racialists or racists, as long as they do not advocate unacceptable policies.

I hope that this contribution will help to de-dramatize the debate, which needs it, to more effectively combat racial discrimination and racism, which are more virulent than ever, and also to prevent the progress of biology and medicine from being hampered, as this is the main danger of an errant "scientific antiracism." My ambition is to inspire a renewed antiracism founded on a concept I believe to be original: the reciprocal sanctuarization between biology, on the one hand, and morality, ethics, and politics, on the other (Tibayrenc, 2019).

Glossary of Specialized Terms

AAI: see: Adaptive Archaic Introgression*.

Adaptive archaic introgression (AAI): Transfer of genes from a primitive (*"archaic"*) *Homo* species (for example: *Homo neandertalensis*) to *H. sapiens*, which contributed to the latter's adaptation to new environments.

AIM: see "ancestry informative marker".

Allele, allelic: Different molecular forms of the same gene. In humans, each chromosome* has two copies. Consequently, a given gene has two copies, one from the mother, the other from the father. If the two copies are genetically identical (have the same DNA* sequence*), the individual is **"homozygous"**. If the two copies are different, the individual is **"heterozygous"** for the gene considered.

Allopatry: Observed in different geographical localities (cf. sympatry*).

Amino acid: Elementary component of proteins*. There are 149 sorts of amino acids in proteins.

Ancestry informative marker (AIM): Genetic markers whose allele frequencies are very different between geographic populations, making it possible to characterize these populations and the individuals belonging to them.

Bacterium: Microscopic unicellular organism, often pathogenic but not always, whose single chromosome* is naked in the cell's cytoplasm, without a nucleus. Example: Koch's bacillus, the agent of tuberculosis.

Balanced selection: Genetic polymorphism that persists in a population because the heterozygotes for the concerned alleles are better adapted than either homozygote. Example: sickle cell anemia* or drepanocytosis (see this term). Homozygotes affected by sickle cell anemia die prematurely, but heterozygotes are protected against malaria*. This genetic trait is common in people of African descent.

Base: Elementary biochemical constituent of DNA and RNA (see these terms).

Base pair (bp): Pair of complementary bases (category of biochemical molecules) in a double-stranded DNA or RNA sequence, consisting of a purine base in one strand linked to a pyrimidine base in the other strand. The base cytosine always pairs with the base guanine, and adenine with thymine (in DNA) or uracil (in RNA) (see: sequence*).

Chagas disease: Parasitic disease prevalent from the southern United States to northern Argentina, caused by a parasite closely related to the agent of sleeping sickness and transmitted by blood-feeding bugs.

Chromosome: a threadlike structure of proteins and nucleic acids found in the nucleus of most living cells, carrying genetic information in the form of genes.

Cline, clinal: When a variation is continuous, without gaps, like the color spectrum.

Clone: Lineages of individuals resulting from asexual reproduction and genetically identical to each other. Observed in certain insects, amphibians, reptiles, and fish with uniparental reproduction ("parthenogenesis") and in many microorganisms. Genetically speaking, in humans, identical twins constitute a clone, as their genome is identical.

Cluster: In evolutionary genetics, this term refers to a visible subdivision within an evolutionary tree constructed from genetic markers (see Figures 3 and 4).

Coding: cf gene.

Common variant: Genetic variant whose frequency is ≥ 5% in a given population. Cf "low-frequency variant" and "rare variant".

Complex disease: Disease caused by an interaction between the environment (lifestyle, for example) and a large number of genes whose individual effect is limited. These genes induce a predisposition to this disease. Example: type 2 diabetes.

Copy-Number Variation (CNV): DNA sequence equal to or greater than 1000 base pairs* (bp), present with a variable number of copies compared to a reference genome. Includes insertions*, inversions*, deletions*, and duplications*. Also called "structural variation*".

Cystic fibrosis: Genetic disease, affecting the glandular epithelia of many organs, particularly the lungs, leading to severe respiratory failure.

Cytokine: Intercellular messenger protein, released by lymphocytes and macrophages. Cytokines promote communication between immune system cells and between them and cells belonging to other categories of tissues. They play an important role in defense against infectious agents.

Deletion: A specific case of mutation in which part of a DNA sequence is lost during replication. Deletions are transmissible to offspring and can have significant phenotypic* consequences.

DNA/RNA sequence, sequencing: cf deoxyribonucleir acid, sequence, sequencing.

Deoxyribonucleic acid (DNA): Substance constituting genes that contains genetic information (see sequence*). Cf nucleotide.

Denisovan: An extinct human group discovered in the Denisova cave, in the Altai Mountains in Siberia. DNA analysis revealed that Denisovans were distinct from Neanderthals but had interbred with them. The taxonomic status of Denisovans is still debated. Denisovans are believed to have lived in the Middle Paleolithic (300,000 to –35,000 years ago).

Drepanocytosis: cf sickle cell anemia.

Duplication: Doubling of a DNA segment (see: Copy number variation).

Electrophoresis: cf isoenzymes.

Endogamy: Situation in which reproduction occurs within genetically related groups or individuals (as opposed to exogamy).

Enzyme: Protein* acting as a catalyst, allowing the level of energy needed for a biochemical reaction to be lowered.

Epigenetics: Activation and deactivation of genes without any change in the corresponding DNA sequence. Epigenetic changes are sometimes hereditary, which goes against the classic belief that only DNA changes are heritable. In other words, parents' experiences, recorded as epigenetic imprints, can be passed on to subsequent generations.

Epistasis, epistatic: When the expression of one gene is affected by the expression of one or more other genes.

Eugenics: Set of selection methods aimed at improving the genetic heritage of a population.

Evolutionary genetics: A field of biology that studies genetic changes (DNA, genes, alleles) within populations and species over time.

Exaptation: Evolutionary process by which a given trait acquires functions for which it was not initially adapted. For example, the forelimbs of the ancestors of bats became wings.

Founder effect: When a population is founded by a limited number of individuals from a larger source population, a large part of the total genetic variability of the source population is lost. Populations resulting from such a founder effect will have reduced genetic variability compared to the source population, and by random sampling, certain alleles that were rare in the source population may be more common in the founded population. For example, French Canadians were founded by a relatively small number of French emigrants. Certain pathological traits (oculopharyngeal muscular dystrophy, tyrosinemia) are more frequent among Quebecers than among the French. The genetic consequences are similar to those of a "bottleneck effect" (when a population passes through a phase where only a restricted number of individuals survive).

Frequency: The number of units in the category under study divided by the size* (see this term) of the population. Example: in a deck of cards, the frequency (expectation, probability) of the category "black card" is ½ = 0.5, = 50%, whether the deck is 36 or 52 cards. On the other hand, the frequency of the category "king" is 4/36 = 0.11 in a 36-card deck, and 4/52 = 0.077 in a 52-card deck.

F**st statistic:** Method proposed by Wright (1951) to describe the genetic structure of a population with three F-statistics: *Fis*, *Fit*, and *F*st. To summarize, *F*st is a measure of genetic differences between two populations, for example, between individuals of European and African descent. Its value ranges from 0 (genetically identical populations) to 1 (populations that are completely different in their genetic makeup).

Gene: Defined portion of the DNA* molecule coding for a given protein* (directing the synthesis of said protein). By extension: any determined DNA segment, whether coding or not.

Gene regulation: Mechanisms implemented to dose the level of gene expression and achieve the optimal expression, in a given cell, of the products of that gene (RNA or protein). Gene regulation explains that, despite all possessing the same genome, the cells of an individual do not express it in the same way: liver cells are very different from cardiac cells. Furthermore, gene regulation varies over time and depends on the environment. It seems to play a considerable role in phenotypic differentiation between populations and species.

Genetics: Study of individual genes.

Genetic distance: Evolutionary difference between two given populations or species, measured with genetic or genomic markers. The genetic distance between humans and chimpanzees, their closest relative, is much lower than the genetic distance between humans and mice, for example. It is of the same order of magnitude as the genetic distance between *Drosophila* (fruit flies) subspecies.

Genetic drift: The evolution of a population or species caused by random events. From a genetic perspective, it corresponds to changes in the frequency of alleles or genotypes, independent of migration, natural selection, and mutations. The effects of genetic drift are greater the smaller the population considered, as this increases the effects of random events on the population, by a lottery effect.

Genome-wide association study (GWAS): Compares the distribution of DNA markers (either microsatellites or SNPs) across the genome in people with a disease or other phenotypic traits compared to people without the disease or these traits. The objective is to identify the genomic region where the genes responsible for the studied characteristics are localized.

Genomics: Study of the entire genome.

Genotype: Genetic constitution of an individual; reading of one, several, or all of their genes.

Geographical race: cf subspecies.

GWAS: see Genome-wide association studies.

Haematophagous: That feeds on blood.

Hard sweep: When strong selection rapidly leads to the fixation of new mutations in given populations (see "soft sweep*").

Hardy-Weinberg equilibrium: population genetics statistic used to verify that in a given population, genetic exchanges occur randomly, in a randomized manner, without obstacles; the population is said to be "panmictic*".

Hemoglobin: Pigment contained in red blood cells, enabling oxygen transport.

Hereditary hemochromatosis: Genetic disease due to excessive intestinal absorption of iron, leading to the deposition of this element in various organs such as the skin, heart, and liver.

Heritability: Proportion of the variability of a given phenotypic trait explainable by the action of genes.

HLA: see Human Leucocyte Antigens*.

Heterozygote: see "allele".

Homogamous, homogamy: Tendency to choose a mating partner with a similar genotype or phenotype.

Homologous chromosome: chromosomes* that have the same shape, the same size and the same genetic content.

Homozygote, homozygous: see "allele".

Human Leucocyte Antigens (HLA*): Cell surface proteins, which exhibit considerable diversity in humans and determine the individual's leucocyte type. The set of HLA genes directs the presentation of antigenic peptides* to the immune system and plays a major role in infectious processes, immunological defenses, and tissue compatibility.

Indel: cf insertion* and deletion*.

Insertion: Additional DNA sequence that inserts into a pre-existing sequence.

Intelligence quotient (IQ): Psychometric test aimed at measuring a subject's general intelligence relative to the average population.

Introgression: Transfer of genetic characteristics from one species to another following hybridization.

Inversion: Mutation causing a change in the orientation of a DNA segment.

IQ: cf intelligence quotient.

Isoenzymes: Protein extracts (enzymes* in this case) from given biological samples, separated by electrophoresis (a method used to separate biological samples on gels or plates by the action of an electric field). These proteins migrate on the plate to a distance that is a direct reflection of the DNA sequence of the genes that determine their synthesis. Isoenzyme variability is therefore a reliable, though coarse, measure of the genetic variability of the studied organisms. From the 1970s, they were a basic method of population genetics for a very long time.

Linnaean nomenclature: (named after the Swedish naturalist Carl Linnaeus, 1707–1778, its inventor). Each species is described with its genus name, then its species name. Humans are *Homo* (genus) *sapiens* (species). Modern humans are the only species in the genus *Homo*. This system is still in use and is applied to all living things: animals, plants, bacteria...

Locus (plural: loci): Physical location of a given gene on the chromosome*. By extension, in genetic jargon: the gene itself.

Low-frequency variant: Genetic variant whose frequency ranges from 0.5% to 5% in a given population (= "Goldilocks variants"). Cf "common variant" and "rare variant".

Malaria: Parasitic disease caused by a protozoan (unicellular parasite) of the genus *Plasmodium*, transmitted by mosquitoes of the genus *Anopheles*. The parasite causes the patient's red blood cells to burst, leading to more or less severe anemia.

Mendelian population: Population within which genetic exchanges obey Mendelian laws (named after the Austrian monk Gregor Mendel [1868–1884]). These Mendelian laws establish the rules of current genetics.

Mesolithic: period of Prehistory chronologically and culturally intermediate between the Paleolithic* which precedes it and the Neolithic* which succeeds it.

Microsatellite (= short tandem repeats; STR): Short DNA sequence, generally 1 to 4 base pairs* (bp*) long, repeated along the DNA molecule. In humans, as in many other species, there is great variation in the number of repeats from one individual to another (widely used in forensic applications for individual identification) and between different populations. The number of repeats for a given locus defines the microsatellite alleles*. There are hundreds of places in human DNA and in most other species that contain microsatellites. To summarize: microsatellites are high-resolution genetic markers (higher resolution than isoenzymes*, but lower than SNPs*), used to characterize individuals and populations.

Mitochondrion: Mitochondria are membrane-bound organelles present in most eukaryotic cells. They are often called the "powerhouses of the cell" because they generate adenosine triphosphate (ATP), the cell's main energy source, through a process called oxidative phosphorylation. Mitochondria have their own DNA (mtDNA), inherited from the mother and coding for some of the proteins essential for their function. In addition to energy production, mitochondria play a key role in cell signaling, apoptosis (programmed cell death), and metabolism.

Mutation: Any change in the structure of a gene, or of any DNA segment.

Natural selection: One of the driving mechanisms of species evolution. It is the differential reproductive success between individuals of the same species and the differential success of genes in a given population, caused by the presence or absence of favorable or unfavorable genetic variations, in the face of a changing environment.

$Neanderthal, Neanderthal man: Archaic species that lived in Eurasia, taking its name from the "Neanderthal" valley in Germany, where the first fossil was discovered in 1856. The oldest known Neanderthals are believed to have lived 430,000 years ago. Neanderthal man disappeared 30,000 years ago. He hybridized with modern humans several times.

Neolithic: period of prehistory and ancient history ranging from -10000 to -2000 years.

Nomenclature: Classification system serving as a reference framework within a given discipline. Linnaean nomenclature concerns zoology and botany.

Nucleic acid: cf deoxyribonucleic acid (DNA) and ribonucleic acid (RNA).

Nucleotide: Organic molecule including, among other things, a nitrogenous base or nucleic base*. Nucleotides constitute the basic element of a nucleic acid* such as DNA or RNA. The percentage of nucleotide differences between individuals, populations, or species is a reliable measure of "genetic distance".

Nucleotide polymorphism: cf Single nucleotide polymorphism.

Paleolithic: prehistoric period from –3 million to –11700 years ago.

Pangenomic: cf genome-wide association study.

Panmixia, panmictic: Said of a population within which genetic exchanges occur strictly at random (in a *"randomized"* way).

Parasitic protozoan: Unicellular pathogen, whose genetic material is contained in a nucleus (eukaryote). Example: *Plasmodium falciparum*, agent of the most severe form of malaria.

Pathogenic fungus: Infectious agent from the fungus group, generally microscopic, responsible for serious pathogenic conditions in humans. For example: *Aspergillus fumigatus*, agent of aspergillosis.

Pathogenic yeast: Specific category of microscopic fungus. Example: *Candida albicans*, agent of candidiasis.

Phenotype: All the observable characteristics of a given individual, population, or species, except the genome (which is the genotype). The phenotype is not limited to morphological characteristics and includes, for example, physiological parameters (blood pressure, muscle strength, etc.), biochemical parameters (cholesterol level, etc.), or cognitive properties. The phenotype is produced by the interaction between the genotype and the environment. The variable part of the phenotype due specifically to the action of the genotype constitutes the heritability (see this term) of the studied trait.

Phenotypic expression: Process by which a phenotype (for example: skin color, or height) is expressed from the specific genes that control it. Phenotypic expression can be largely modulated by gene regulation* and epigenetics* in the absence of any modification of the concerned DNA sequences.

Phenylketonuria (PKU): Rare genetic disease causing mental impairment, linked to a deficiency of the enzyme phenylalanine hydroxylase, leading to the accumulation of the amino acid phenylalanine in the brain and blood. It is a curable disease by a low-phenylalanine diet, which allows for normal or at least improved intellectual development.

Phylogenetics: Branch of genetics that aims to reconstruct the evolutionary past and genetic relationships of separated taxa or evolutionary lineages. Phylogenetics reconstructs the genealogies of living beings. These genealogies are visualized by "phylogenetic trees" developed from specialized computer programs (see Figures 3–7 in Part I).

Phylogenetic signal: When a set of genetic or genomic data in a given population allows for the identification of clear discontinuities (clusters) within this population. In the absence of such a signal, genetic variability appears as an undifferentiated continuum. This is what is observed in a panmictic* population.

Pleistocene: First geological epoch of the Quaternary period. It extends from –2.58 million years ago to –11,700 years before the present time.

Point mutation: Mutation that affects only a single nucleotide* of a nucleic acid*.

Polygenic: Relating to several genes.

Polygenic Risk Score (PRS): In genetics, a Polygenic Risk Score, also called a polygenic score, genetic risk score, or genome-wide score, is a number that summarizes the estimated effect of many genetic variants on a given phenotype of a given individual, for example, the risk of developing a given disease.

Polypeptide: Organic molecule composed of a chain of more than 4 amino acids*.

Popper's falsifiability criterion: cf Popper.

***Popper (falsifiability criterion of), Popperian:** (from the philosopher Karl Popper, 1902–1994) A Popperian scientific approach considers that a scientific hypothesis is more informative the more it is falsifiable. It favors a *deductive* approach (one seeks to refute the hypothesis one wants to support) over an *inductive* approach (one seeks to accumulate facts favoring the hypothesis).

Population genetics: Analysis of genetic variation in space and time within and between populations. Population genetics focuses on the population or species rather than the individual. Cf. population genomics*.

Population genomics: Analysis of population variability based on the study of the entire genome.

Protein: Biological macromolecules formed from one or more polypeptide* chains. Each of these chains consists of a sequence of amino acid* residues.

Proteomics: Refers to the science that studies proteomes, i.e., the entire set of proteins in a cell.

Randomized: Statistical study in which subjects are drawn at random.

Rare variant: Genetic variant whose frequency is less than 0.5% in a given population. Cf "low-frequency variant" and "common variant".

Ribonucleic acid (RNA): A nucleic acid chemically very close to DNA. Several sorts of RNA exist. Messenger RNA (mRNA) and transfer RNA (tRNA) transmit the genetic information from DNA to direct protein synthesis. Cf nucleotide.

Scientific anti-racism: Concept according to which human races have no biological reality and are purely sociocultural concepts, which renders racism inept.

Self-reported ancestry (SRA): Protocol for studying human populations in which individuals themselves identify their ancestry group, generally from a pre-established list proposed by the study's author.

Sequence,sequencing: DNA/RNA sequence can be compared to a series of letters representing the primary structure of an actual or hypothetical DNA or RNA molecule or strand. The possible letters are A, C, G, and T, which correspond to the four nucleotide subunits of a DNA strand (the bases* adenine, cytosine, guanine, thymine; see: "base pair*"). This coded sequence represents the basic genetic information. A DNA sequence can code for proteins, via messenger and transfer RNAs. In this case, the DNA sequence directly controls the succession of amino acids that constitute the primary structure of the protein. Some other sequences have no known coding function.

Sexual selection: Selection that favors traits (e.g., the peacock's tail) that increase mating success. Sexual selection, if it goes too far, can counteract general natural selection and can be detrimental to the survival of the individual and the species.

Short tandem repeat (STR): cf microsatellites.

Sickle cell anemia (= drepanocytosis): A genetic condition common in individuals of African descent. It affects hemoglobin, the protein that transports oxygen in red blood cells.

Size: the number of units that make up a population. Example: in a deck of cards, the population size is either 36 or 52.

Single-nucleotide polymorphism (SNP), single-nucleotide variant (SNV): Polymorphisms of a single letter variation (a single point mutation) in the DNA sequence. SNPs contribute to differences between individuals and populations. Most of them have no phenotypic effect, others cause subtle differences in countless characteristics, such as appearance, while some affect the probability of certain diseases occurring. SNPs are widely used as high-resolution population markers. It is important to note that there are variations in allele* frequencies between different human populations, so an SNP that is common in one geographic or ethnic group may be much rarer in another.

SNP, SNV: cf single-nucleotide polymorphism, single-nucleotide variant.

Sociobiology: Discipline aiming to explore the biological and genetic foundations of animal and human behaviors.

Soft sweep: Selection acting on a large number of genes already present ("standing variation") whose individual effect is limited, and on epistatic regulatory phenomena (see "hard sweep*").

Species: Taxonomic level to which modern humans belong. The species is defined according to varied criteria. A given species receives a Latin name (biname) composed of the genus name (taxonomic category above species) and the name of the species concerned. Humans, thus, are *Homo sapiens*. For the different definitions of species, see text.

SRA: cf self-reported ancestry.

Standing variation: Genetic variation already present in a given population.

Strain: In microbiology, refers to lineages of microorganisms that exhibit common properties. Examples: a virulent strain, a penicillin-resistant strain.

Structural variation: cf copy-number variation.

Subspecies (also called geographical race): Geographical populations of the same species that exhibit specific phenotypic traits making it possible to distinguish most individuals of one population from most individuals of another population. Subspecies can also be defined by different genetic frequencies.

Sympatry: Said of individuals or populations sharing the same geographical area (cf "allopatry"*).

Synthetic theory of evolution (= Modern evolutionary synthesis): Synthesis of Charles Darwin's theory of evolution* with later contributions from Mendelian genetics and population genetics which were unknown to Darwin. Also called "neo-Darwinism", it was proposed before the last war by, notably, R. A. Fisher, J. B. S. Haldane, Sewall Wright, Theodosius Dobzhansky, Julian Huxley, Ernst Mayr, Bernhard Rensch, George Gaylord Simpson, and George Ledyard Stebbins. Cf theory of evolution.

Systematics: cf taxomomy.

Taxonomy: Science of classifying living beings.

Theory of evolution: Theory developed by the English naturalist Charles Darwin in his work "On the Origin of Species" published in 1859. It posits that species evolve and derive from one another through natural selection. Cf synthetic theory of evolution*.

Transcription: cf transcription factor.

Transcription factor: protein that is essential in the process of genetic regulation*, which binds to specific genes and modulates the rate of transcription of genetic information, transcription being the mechanism that allows DNA to be encoded into messenger RNA. Messenger RNA is then encoded into proteins by the phenomenon of translation.

Transcriptomics: the complete set of RNA molecules produced by a cell.

Translation: cf transcription factor.

Variance: Sum of the squares of the deviations from the mean. More simply: a measure of the variability of a given trait.

Virus: Infectious agent generally composed of a nucleic acid molecule wrapped in a protein envelope. A virus is too small to be seen under light microscopy and is only capable of multiplying in the living cells of its host.

WGS: cf whole genome sequencing.

Whole genome sequencing: sequencing of the whole genome, the fundamental method of the discipline called "genomics".

Part I

The Biological Race

Chapter 1

The Debate

"Scientific Antiracism"

Scientific antiracism is the strategy aimed at draining racism of all substance by scientifically denying any biological reality to human races. I am stating my psosition clearly, and this is a major theme of this book: I do not support this strategy, which I find debatable on a scientific level. I believe it largely boils down to a semantic quarrel and that it is counterproductive from a societal point of view, whatever the highly honorable nature of the intended goal.

What is Racism?

Before engaging in the debate, it is more than useful to define unambiguously this term, which is currently applied to everything, and which is often simply a semantic weapon to disqualify a political opponent when arguments are lacking. Calling someone a "racist," often coupled with "fascist," acts as a political, media, and social excommunication, or, to use the current buzzword, "demonization." Moreover, the word has become excessively generalized and diluted with a host of derivative meanings: "anti-cop racism," "anti-bourgeois racism," "anti-youth racism," etc.

So, what is meant by racism—at least by me, in this book? Here is the definition given by the dictionary of the French academy: "a set of doctrines according to which the varieties of the human species called races, principally distinguished from one another by their physical appearance, would be endowed with unequal intellectual and moral faculties, directly linked to their genetic heritage."

Racism is therefore the approach that consists of attributing intrinsic, innate, biological properties to the geographical populations that make up humankind (for example: original inhabitants of Western Europe, intertropical Africa, Southeast Asia), particularly in the cognitive and behavioral domain, and, more specifically, in deducing a *hierarchization* of these populations (Anonymous, 2024a): humankind is thereby divided into "superior races" and "inferior races." In a more advanced form, racism consists of drawing political and historical conclusions from this supposed state of affairs, with the "superior races" having a natural vocation to dominate the

DOI: 10.1201/9781003647393-2

others, or even to "civilize" them. In the present book, only this meaning, which implies a *hierarchization* of geographical populations, will be retained.

> *The strict sense of racism implies a* <u>*hierarchization*</u> *of human geographical populations, particularly by assuming they possess different innate cognitive properties.*

According to this precise meaning, racism must therefore be distinguished from racialism (the approach consisting of recognizing the existence of human races without the tendency to hierarchize them), ethnicism (the observation of cultural differences between geographical populations), and ethnocentrism (the conviction that one's own ethnic group is superior to others culturally) (Takezawa et al., 2023). I will return to these differences between race and ethnicity.

Regarding the act of distinguishing races without hierarchizing them, it should be noted that Ruth Benedict, a student of the anthropologist Franz Boas, from the pre-war antiracist movement in the United States, clearly distinguished between racism on one hand, and the study of races on the other hand. We can also recall the quote from the former French minister Alain Peyrefitte, who denounced "the conformism according to which, to admit the plurality of human races is to approve of the crematoria" (de Benoist, 1979).

Finally, we can cite the first UNESCO statement (1951) on racism (UNESCO statement on race, 2022): "The physical anthropologist [...] knows that races exist [...] from the scientifically recognizable and measurable congeries of traits which he uses in classifying the varieties of man." In this initial statement, unlike the subsequent ones, UNESCO did not question the existence of human races but merely condemned racism (Sesardic, 2013).

In this book, in accordance with my area of expertise, I particularly analyze the problem of human races and of racism or antiracism *on the biological level.* In France, the phenomena of racism or ethnic division largely concern cultural, religious, economic, historical, and political aspects: the Algerian war continues to haunt minds, as does the Israel/Palestine conflict. All of this has little to do with biology.

Biological racism is what Charles Maurras, the leader of the "Action française" (French Action) political movement, at the beginning of the XX° century, contemptuously called "skin racism." This is the kind that, for example, motivates the "identitarians" in France, for whom a "true" Frenchman cannot be a "racialized" person—Black, Arab, or Asian. To be a "true" Frenchman, one must "show a clean bill of health" (or "show a white paw"). This is similar to the "white supremacist" movement in the USA.

Essential Biological and Taxonomic* Reminder

> ***Taxonomy: "Nature laughs at our classifications"***
> ***French Société polytechnique, 1827***
>
> *"Nature, although inexhaustible in its combinations, does not however proceed by leaps and bounds; on the contrary, it is through transitions subtly managed that it incessantly spins the immense chain of organized beings; it is by links of imperceptible texture that it binds all that extends, vegetates, or breathes. We establish classifications and methods with great difficulty and art, and it laughs at our methods and our classifications."*

Biological racism addresses man as a biological being. It is therefore useful to present a reminder to situate man in the living world.

> *Taxonomy is the discipline that allows for an essential classification of living beings.*

To bring order to the profusion of past and present living beings, scientists establish taxonomies, which are based on the hierarchical KPCOFGS system, for Kingdom, Phylum, Class, Order, Family, Genus, Species, in descending order.

In the first instance, these different categories do not have a clear definition, apart from their rank: thus, a class contains several orders, an order several families, and so on. In many cases, to improve the precision of the classification, scientists add supplementary categories: tribe between genus and family, super-order above order, species complex above species, subspecies below, etc. It is a system of Russian dolls.

Only the species has been the subject of (numerous) attempts at precise biological definitions. According to the KPCOFGS classification system, man can be classified as follows: kingdom Animalia, phylum Chordata, class Mammalia, order Primates, family Hominidae, genus *Homo*, species *sapiens*. *Homo sapiens*, the "wise man" (no comment). *H. sapiens* is the only existing species of the genus *Homo*. In addition to *H. sapiens*, the Hominidae family also includes the chimpanzee (*Pan troglodytes*), the bonobo (*P. paniscus*), the gorilla (two species: *Gorilla gorilla* and *G. beringei*), and the orangutan (two species: *Pongo pygmaeus* and *P. abelii*).

The Invaluable Contribution of Evolutionary Genetics and Population Genetics

This book focuses on the *biological* nature of human races, emphasizing the contribution of *evolutionary genetics* and *population genetics*. These disciplines, which are my domain of expertise, reconstruct the evolution of populations and species using the tools of genetics. They provide highly reliable insights into the genetic characteristics of human geographic populations, as well as their kinship relationships and the migrations that shaped them. In contrast, the never-ending controversies over the comparative cognitive abilities of human geographic populations provide us with infinitely less reliable data (see: "recent evolution of cognitive traits").

In applied research (exploration of genes and phenotypes of interest), it is advisable to begin by establishing, *before any study*, a general genetic framework for the species studied throughout its ecogeographical range. The genes and phenotypes of interest are *then* mapped onto this general framework. This allows the specific evolution of these genes and phenotypes to be compared with the overall evolution of the entire species (phylogenetic character mapping; Avise, 2004). This approach is highly recommended for applied research in humans.

This approach to developing an upstream genetic framework constitutes the most robust part of the argumentation developed in this book.

The Species Concept Puzzle

To the non-specialist, the species seems like a simple and well-defined concept. A tiger is different from a lion, a dog from a wolf, a polar bear from a brown bear. Or is it? The last two examples were deliberately chosen to precisely pinpoint the roots of the puzzle.

> *Contrary to what one might think, the species concept does not have a strict, unique definition.*

The most widespread species concept is the biological, or mixiological, criterion: a species is a reproductive community (Dobzhansky, 1937). Thus, all members of the species *Homo sapiens* are potentially interbreeding. The only barriers to this within our species are geographical and cultural ones. In contrast, the donkey and the horse are different species because, although they can produce offspring (mules, hinnies, depending on the direction of the cross), these hybrids are sterile. There is therefore an interruption of gene flow. Fine, but it is not always so simple. For example, the wolf (*Canis lupus*) and the coyote (*Canis latrans*), though classified as different species, can perfectly hybridize, and the hybrids ("coywolf") are fertile. The same is true for the polar bear (*Ursus maritimus*) and the North American brown bear or grizzly (*Ursus arctos horribilis*), whose hybrids ("pizzly") are also fertile. Due to global warming, the territories of these two species are increasingly overlapping, which favors these trans-species romances.

Apart from the mixiological concept, two other species concepts are commonly used: the phenotypic* concept and the phylogenetic concept.

The phenotypic concept is still widely used today. Species are described based on their external appearance (phenotype). Indeed, very few descriptions of new species are based on verifying that they are genetically isolated from other species (the mixiological criterion).

The phylogenetic concept of the species (Cracraft, 1983) is based on the principle that species correspond to evolutionary lineages distinct from one another, which can be individuated by phylogenetic analysis and their hierarchical relationships, visualized in the form of evolutionary trees, or genetic genealogies.

Unfortunately, the problem of defining the species is not limited to these three concepts. In fact, there is no consensus among specialists on what a species should be. For instance, Hey (2001) cataloged no fewer than 24 different species concepts!

> *"A species is what an experienced and competent specialist considers to be a species."*

To introduce a little humor into this dry debate, I will quote the above definition, whose source I have forgotten.

The somewhat arbitrary nature of the species concept was recently illustrated by the supposed "de-extinction" of the Paleolithic* giant wolf, the "dire wolf" (*Canis dirus*), which disappeared 12,000 years ago (Lamoureux, 2025). The American company "Colossal Biosciences®", which is also seeking to resurrect the mammoth and the Tasmanian wolf, announced the de-extinction of this giant wolf. In fact, this resurrection was limited to inserting a few dozen key genes into the genome of cells from the current wolf, which is supposed to confer on the individuals born from this manipulation the essential characteristics of *Canis dirus*: large size, white color, powerful jaws, and blood-curdling howls. But does this make it a "real" *Canis dirus*? Can we say that the wolf pups born this way belong to the species *Canis dirus*? Yes, according to the company, which announces: "if they look like the animal, then they are the animal." This is highly debatable, even if one can salute the technical achievement.

Fortunately, to simplify the question within the context of the present book, our species, *Homo sapiens*, turns out to be a "good" species according to the three main criteria presented above.

- Mixiological: All members of the *H. sapiens* species are potentially interbreeding, and there is strict genetic isolation from other species, even the most closely related ones (chimpanzees and bonobos).
- Phenotypic: Even the most inept zoologist is capable of identifying our species on morphological grounds, despite its extreme variability (Sarich and Miele, 2004), and of distinguishing it from any other species.
- Phylogenetic: *H. sapiens* constitutes a well-individuated evolutionary lineage, clearly distinct from other species in the hominid family.

The Subspecies or Geographical Race: A Highly Relevant Concept for This Debate*

A subspecies corresponds to a geographical subunit of a given species exhibiting certain remarkable traits that allow it to be reliably identified from other subspecies of the same species. The Dictionary of Science and Technology by Larousse (Aitken et al., 1995) gives the following definition: "taxonomic subdivision of a species, exhibiting certain morphological differences compared to other subspecies, and often a different geographical distribution and ecology." The Dictionary of Genetics by King, Mulligan and Stanfield (2012) gives a similar definition: "(1). A taxonomically recognizable subdivision of a species; (2). Geographically and/or ecologically defined subdivisions of a species, with distinct characteristics." These authors refer to the term "race." Other authors add a genetic component to the definition of the term (see: "The concept of subspecies or geographical race = ancestry groups.").

The recommended limit for being able to describe subspecies is the possibility of distinguishing at least 75% of the members of a subspecies from 100% of the individuals of all other subspecies of the same species (Adamon, 1949). Subspecies are a well-known entity among zoologists. With their infinite varieties, they are the delight of amateur entomologists who collect flower chafers and carabid beetles.

The subspecies concept (geographical subdivisions of the same species identifiable by certain specific characteristics) is very important for the present debate.

Due to Linnaean terminology* (from the Swedish naturalist Carl Linnaeus, 1707–1778; see below: "history of the notions of race and racism"), subspecies are given a triname: the name of the genus, the name of the main species, and the name of the subspecies. Thus, the European ox and its exotic cousin, the zebu, are named respectively *Bos taurus taurus* and *Bos taurus indicus* (Figure 1).

It should be noted that the European ox and the zebu were formerly cataloged as different species (*Bos taurus* and *Bos indicus*). This illustrates the element of arbitrariness that mars all taxonomy. Arbitrariness does not mean "meaningless." The taxonomic effort is essential to bring order to the profusion of the living world and to delimit the units of analysis that will be the subject of scientific studies.

Subspecies are generally described based on morphological traits ("phenotypic" subspecies). Thus, the European ox and the zebu exhibit obvious morphological differences: the zebu has a dorsal hump, drooping ears, and a large skin fold on its chest. These differences are not limited to external appearance. Subspecies, in general, can be distinguished by different allelic* frequencies*. They may differ in

Figure 1: Two subspecies of domestic animals. On the left, the European ox (*Bos taurus taurus*); on the right, the zebu (*Bos taurus indicus*; Thailandese specimen).

their susceptibility to diseases. For example, the European ox is highly susceptible to a parasite prevalent in many tropical regions, *Trypanosoma evansi*, which is a cousin of the agent of sleeping sickness in Africa. In contrast, the zebu, due to its millennia-long exposure to this parasite, has become genetically resistant to it. Thailandese veterinarians have utilized this property to create lineages (a little zebu, a lot of Dutch cow) resistant to the parasite but maintaining good milk production.

Indeed, by definition, individuals of different subspecies are perfectly interbreeding. This is the case for the domestic ox and the zebu. Many individuals exhibiting all degrees of mixing between these two subspecies exist.

Populations showing the beginning of genetic isolation are called "nascent species" (species in statu nascendi) or "semi-species" (Dobzhansky and Spassky, 1959). Finally, let us recall that certain populations considered different species remain interbreeding, with fertile hybrids. This is the case for (1) the wolf (*Canis lupus*) and the coyote (*C. latrans*); (2) the North American brown bear or grizzly (*Ursus arctos horribilis*) and the polar bear (*U. maritimus*). Yes, truly, nature laughs at our classifications! They are nonetheless indispensable. Which matters is to be aware of their limits.

An important remark concerning the definition of subspecies: it is a concept that relates to *populations*. Subspecies are described according to characteristics that concern all the individuals that compose them. There will always be a few atypical individuals who seem to stand out. One must not focus on these few exceptions to grasp the subspecies concept. This remark is of major importance when considering human geographical populations: they must be analyzed with a *population-based* perspective.

This essential reminder completed, let us return to our debate.

Brief History of the Concepts of Race and Racism

Contrary to a widely held doxa, our ancestors were not "color blind" and perfectly distinguished the different peoples they encountered based on their physical appearance (Sarich and Miele, 2004).

> *Peoples of antiquity were by no means "color-blind" and distinguished geographical populations by their exterior aspect.*

The ancient Egyptians distinguished the Rits or Egyptians, painted in red, the Namous, with aquiline noses, painted in yellow, the Nashu, black with kinky hair, and the Tenehous, blond, blue-eyed Berbers (Anonymous, 2022a). It is surmised that these distinctions were not accompanied by any hierarchization. What is certain is that the ancient Romans took people of all origins as slaves, including Europeans.

> *The significant development of the natural sciences in the 18th century ("Age of Enlightenment") went hand in hand with the advent of racism, colonialism, and slavery.*

The 18th century, with the advent of the Enlightenment, was accompanied by a burgeoning development of the natural sciences; it was the great era of Linnaeus, Buffon, Lamarck, Daubenton, von Humboldt. A scientific, one might say "zoological," vision of human diversity then developed. The different human populations were defined as taxa, with scientific names in Latin.

Carl Linnaeus (1707–1778), a Swedish naturalist, is the inventor of the binomial nomenclature, also called Linnaean nomenclature*. Each taxon receives a genus name, followed by the species name, both in Latin. Thus, the scientific name of modern man is *Homo sapiens*, with *Homo* being the genus name and *sapiens* the species name. In the case of a subspecies, a triname is used. This binomial or trinomial nomenclature, as we have seen (see: "biological and taxonomic reminder"), is still in use today, for animals as well as for plants and bacteria.

Linnaeus undertook the classification of different human populations and assigned them trinames, thereby effectively making them subspecies: *Homo sapiens europeus*, *asiaticus*, *afer* (African), *americanus* (Native American).

Johann Friedrich Blumenbach (1752–1840) was a German physician, anthropologist, and biologist. He classified humanity into five races: Caucasian (European), Mongol (Asian), Malay, Black (African), Red (Native American).

Contemporary Era: Racism and Politics

Racism has powerfully served an agenda of domination of certain peoples by others. It has been repeatedly invoked to justify slavery, colonialism, and segregation. One of my favorite books, which I have owned since childhood, is a luxurious popular science work on zoology, "les animaux" (Animals) (Joubin and Robin, 1923), which dates from 1923. The chapters devoted to animals are remarkable and show that zoological science, a century ago, was already quite advanced. But the chapter dealing with the "king of creation," meaning Man, is quite simply astonishing for a contemporary mind. It is steeped in a quiet, assumed, monolithic racism. Central importance is given to the "facial angle," supposed to be correlated with brain development, and consequently, intelligence (Figure 2). A few gems follow, such as: "it follows that the white man is very clearly the most intelligent; he also dominates the earth; he colonizes worldwide, and maintains the other races in servitude, not for their good, alas! but for his own interest." One can appreciate this embryo of repentance.

To the credit of the authors, L. Joubin and Aug. Robin, both highly distinguished scientists, this book is only a faithful reflection of the ideas of the time. At that time, which saw the heyday of colonial expansion, the "superiority" of the white man (preferably Anglo-Saxon) seemed indisputable, and his domination over other peoples, completely natural. If one has the curiosity to read the short story "Allouma," by the 19th Century French writer Guy de Maupassant, recounting the love affair between a French colonist and a nomadic Algerian girl during the Second Empire, notations abound that are totally inaudible in our era: "… for those creatures, on African soil, belonged to us almost body and soul."; "I immediately felt that her fleeting, savage thought had been extinguished in rest"; this embellished with sexist reflections on "the impure gaze of women," etc. We are of course far from the *Woke culture*.

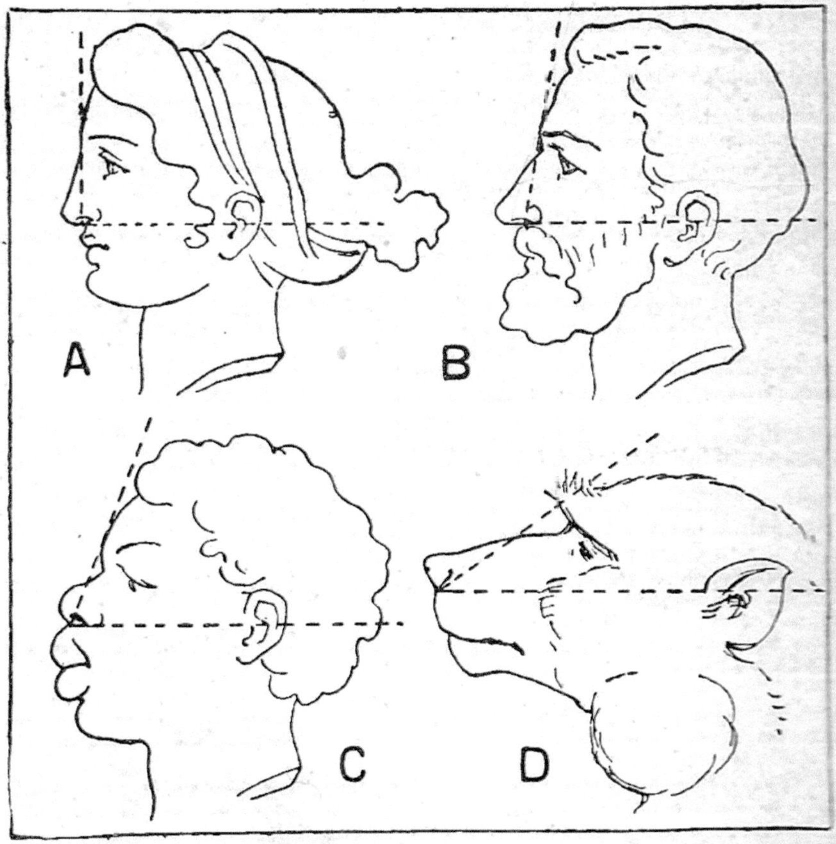

Figure 2: Comparative measurement of the facial angle in A: the Ancient Greek; B: the European; C: the "Negro"; D: the monkey, probably a baboon. Based on Joubin and Robin, 1923.

Albert Einstein, the famous physicist, Nobel Prize winner, is usually considered an indisputable moral authority. It is true that in the 1940s, he passionately defended the cause of African Americans, to the point of being placed under FBI surveillance. However, reading his travel notes, written in 1922–1923 (he was 43 at the time) reveals a darker side of the man (Vey, 2018). These writings are full of decidedly racist reflections on the Chinese and the inhabitants of Sri Lanka. Only the Japanese found favor in his eyes. Einstein›s case is yet another illustration of the fact that in the 19th and 20th centuries until the Second World War, racism was a system of thought widely prevalent in very diverse circles, including among individuals who could not be considered fools, such as the authors of the book "les animaux," (Animals), Maupassant, or Einstein.

Until the last World War, racism was a dominant thought in Europe and North America, including among the elites and left-wing politicians.

Contrary to a common idea, racism—at least the concept we currently have of it—was not the prerogative of one political side. Colonialism, eminently supported by racist ideology, was promoted in France not by the Right, but by the Left, the Jules Ferry chamber in the National Assembly (Girardet, 1972). I am speaking here of the 2nd colonial expansion, that of the 19th century. During a memorable speech to the French National Assembly on July 28, 1885, the creator of secular and compulsory primary education, Jules Ferry, uttered this famous sentence: "There is a right for the superior races, because there is a duty for them. They have the duty to civilize the inferior races." Racism that can be described as benevolent and paternalistic, but undeniable, with an unambiguous hierarchization of the races. Meanwhile, the Right fiercely opposed colonialism, not because the right-wing deputies were kind, but because they saw it as a disastrous dispersion of national energy to the detriment of the only noble goal in their eyes: the reconquest of Alsace-Lorraine: "I have lost two sisters and you offer me twenty servants," thundered the poet Paul Déroulède in the assembly, the "twenty servants" betraying a not-so-benevolent racism. It was only later, in the 20th century, that the Right "turned its coat" on colonialism and saw the interest of preserving the colonial heritage. As for the Left, it persisted and signed onto benevolent racism until a much more recent time. On July 9, 1925, in the National Assembly, Léon Blum (yes, the emblematic leader of the Popular Front!) declared: "we admit the right and even the duty of the superior races to draw to themselves those who have not attained the same degree of culture and to call them to the progress realized thanks to the efforts of science and industry." A definitely different era.

The Rejection of Racism and Eugenics*, UNESCO Statements, and the "Absolute Paradigmatic Shift"

Going against the dominant ideas of the time, certain scientists, particularly in the United States, committed themselves to the path of antiracism starting in the interwar period.

Despite this apparent consensus on the "superiority" of certain races, a powerful intellectual movement had attempted to combat these ideologies—racism and eugenics—which continued to exert great influence on the flow of ideas, even before the war. In the United States, one of the leaders of this movement was the American anthropologist Franz Boas, followed by his students Ruth Benedict and Margaret Mead. It is worth noting that Ruth Benedict, although sharing her mentor's antiracist and egalitarian vision, recommended distinguishing between racism and the study of races.

In the United Kingdom, during the same period, the principal architects of this antiracist movement were the biologists John Burdon Sanderson Haldane, Julian Huxley, and Lionel Penrose. It should be noted that Haldane and Huxley were proponents of a "left-wing" eugenics, achieved through the improvement of living conditions to ensure the success of eugenics. Penrose defended the idea that intelligence resulted from a combination of the effects of environment and genetics; in other words, his view was close to that of current authors such as Hans Eysenck, Arthur R. Jensen, John Philippe Rushton, or Charles Murray, who are considered racist today. This would certainly be the case for many "antiracists" of the pre-war era, who would be consigned to the bench of racist infamy nowadays (UNESCO statement on race, 2022).

After the end of the last war, the debate clarified and polarized. Following the discovery of the atrocities committed by the Nazis, a universal rejection of racism was established. UNESCO issued several declarations to affirm this movement. A little-known fact is that the first two declarations, those of 1950 and 1951, did not deny the existence of human races, and merely rejected any idea of racial superiority and solemnly condemned all discrimination based on race or origin. Four main propositions were formulated: (1) The mental capacities of all races are similar; (2) There is no indication that hybridization (understood as miscegenation) leads to harmful biological consequences; (3) There is no correlation between, on the one hand, given national or religious groups, and on the other hand, given races; (4) Race is a concept more social than biological.

> *After the last world war, UNESCO endeavoured to drain racism of its substance by attempting to demonstrate that "human races do not exist" (the "absolute paradigmatic shift"). This is what is called "<u>scientific antiracism</u>" (disqualifying racism through a <u>scientific</u> approach).*

The Director-General of UNESCO at the time was Julian Huxley (see above). In the past, this author was not exactly part of the antiracist movement, as he is credited with the following note (1924): "The Negro's mind is as different from the White man's mind as the Negro's body is different from the White man's body." It was he who was the first to propose replacing the term "race" with "ethnic group". This euphemisation, while responding to a wish to cleanse this debate of sinister historical impurities, is questionable in more ways than one, as we will see in Chapter 2 of Part II.

The UNESCO committee responsible for drafting the first declaration, that of 1950, was mainly composed of humanities specialists. However, the rapporteur was Ashley Montagu, a biological anthropologist well-known for his unambiguous positions, seen as a "divisive" Figure.

These first versions, particularly the one from 1951, carried confusion, because while they affirmed that the process leading to the appearance of races was unknown (which is more than questionable, as we will see), they admitted that many populations could be classified in racial terms. These ambiguous assertions, which amounted to saying that we did not really know what races were, but that it was possible to classify humanity into races, could only sow great confusion in the public mind and even within the scientific community. It was the subsequent versions (UNESCO statement on race, 2022) that definitively adopted the central dogma of the "absolute paradigmatic shift" (see below), but without managing to close the debate on the existence or non-existence of human races. I mention below certain

Figures of "scientific antiracism," which will give me the opportunity to develop some of the main arguments of this school of thought, of which they were talented defenders.

These debates within UNESCO were punctuated by the successive interventions of the famous French anthropologist Claude Levi-Strauss (Stoczkowski, 2007).

The Ambiguous and Shifting Position of the Famous French Anthropologist Claude Levi-Strauss Caused Trouble within UNESCO.

This author had produced a text for UNESCO in 1952 entitled "Race and History," which seemed to support the antiracist doxa of UNESCO. But a second essay, titled "Race and Culture" (1971), far from advocating the mixing of cultures, defended on the contrary the idea that each culture should maintain a certain autonomy, the price of human creativity. This proposal caused a certain trouble within UNESCO.

Richard C. Lewontin, Paladin of Scientific Antiracism

The "absolute paradigmatic shift" (Anonymous, 2002a) is the concept according to which human races have no concrete biological existence and are merely socio-politico-cultural creations. It was qualified as absolute to emphasize how much it implied a complete upheaval of previously prevailing viewpoints. The message intended to be conveyed, sometimes very subtly, even subliminally, is that the biological differences between races are minimal, superficial, and, quite frankly, negligible—that is to say, they are meant to be *ignored.*

Richard C. Lewontin (1929–2021) was an ardent and talented champion of this concept, of this "scientific antiracism." The man is certainly not a Figure who leaves people indifferent. A Marxist, and one who did not hide it (Berry and Petrov, 2021; Dietrich, 2021), he was a leader of the "science for the people" movement, a left-wing organization stemming from the pacifist movement of the 1970s. Lewontin was always a fierce opponent of any idea of biological determinism. Associated with his accomplice Stephen J. Gould (1941–2002), an eminent paleontologist, he notably leveled virulent criticism against E.O. Wilson's famous book "On Human Nature" (1978), a foundational book of sociobiology*, which he and Gould condemned as "naive and dangerous oversimplification" (Berry and Petrov, 2021). Undeniably, a man of conviction, at the forefront of all the left-wing struggles of the time: not only scientific antiracism, but also economic inequality and the Vietnam War. Lewontin's numerous detractors accuse him of "tainting his science with his politics," while his defenders endow him with the oxymoronic quality of "anti-dogmatic dogmatism" (Berry and Petrov, 2021). However, his critics cannot argue that he was a poor scientist. He was one of the greatest, a pioneer in the use of mathematics in population genetics, and his work is authoritative. So, what arguments did this author develop to try to empty the biological concept of race of its substance?

The Apportionment of Human Diversity: The Fst *Argument**

"The Apportionment of Human Diversity" (1972) is the title of Lewontin's key article in his antiracist struggle. It has become a true standard, a mandatory passage for this movement.

The era in which Lewontin's historical article appeared was one of a genuine revolution in population genetics. For a long time, the only materials available to researchers in this discipline were the numerous visible mutations* in *Drosophila* (fruit flies): red eyes, wrinkled or atrophied wings, etc., and the giant chromosomes of dipteran larvae, all characteristics concerning only the phenotype.

> *Lewontin's seminal work was based on the advent of the first molecular genetic markers.*

The 1970s saw the advent of genetic and molecular markers: blood groups, protein* and isoenzyme* electrophoresis*. Although these traits technically remain phenotypes, their genetic control is perfectly known, which means that they allow for the first time a direct reading of the genotype* and a quantification of genetic variability. Lewontin was a pioneer in the mathematical and statistical treatment of this genetic and molecular polymorphism

I have an in-depth knowledge of these "outdated" markers, as I based a large part of my career on their use (Tibayrenc, 1979, 2009). I am therefore well-placed to judge their value. They have provided invaluable service. However, their resolving power is much lower than current methods: whole genome sequencing (WGS) and single nucleotide polymorphism* (SNP*). In short: the "outdated" markers of the 1970s and 80s are a magnifying glass, while the most modern markers are an electron microscope, with the markers of the 1990s–2000s like microsatellites*, which have an intermediate resolving power, being a light microscope (Figure 3).

What did this famous 1972 paper, still abundantly cited by proponents of "scientific antiracism," say?

> *The largest part of the genetic variability of the human species is found within geographical populations ("races"),*
> *based on "common" genetic variants (frequency ≥2%).*

It stated that the greatest part of human genetic diversity, about 85–95%, was observed *within* the continental geographical populations (the "races"). The differences *between* these populations accounted for only 5–15% of the total genetic diversity of our species. This result, based essentially on the statistic known as *F*st (see Box 1), when taken literally, simply stated that there could be much more genetic difference between two Europeans than between a European and an African. There is

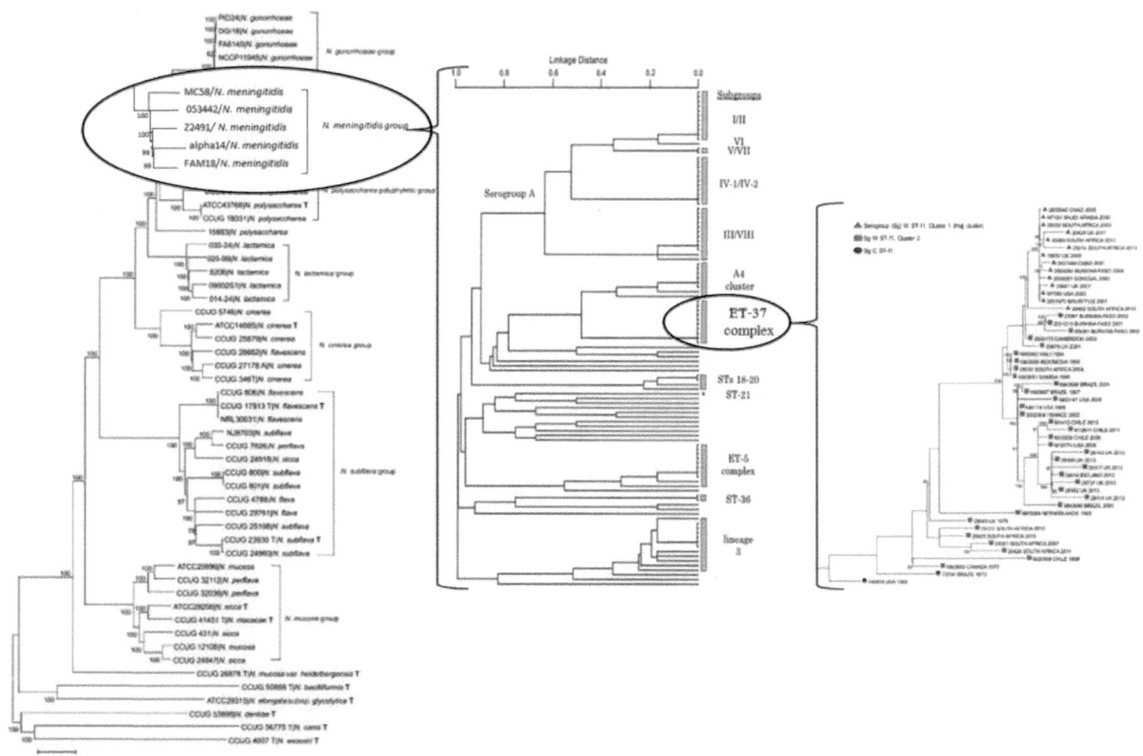

Figure 3: The same biological and evolutionary reality (the genetic diversity of *Neisseria meningitidis*, a bacterium causing meningitis) viewed by three genetic markers of different resolution; Left, low-resolution marker ("magnifying glass"). The group indicated by an oval shows no genetic variability with this marker. Middle, intermediate-resolution marker ("light microscope"): the group marked by an oval on the left figure shows considerable genetic heterogeneity. Right, high-resolution marker ("electron microscope"). The group marked by an oval on the middle figure, which seemed invariant, is no longer invariant at all (Figure author: Dr. Jenny Telleria, IRD Montpellier, France).

no doubt that such a picture seems to considerably weaken a view of human races as profoundly different from one another. The "races," according to Lewontin's results, each appear to be highly diverse genetically, and the genetic differences observed between them "minimal," even "negligible."

This "knockout argument" has been used massively by opponents of the concept of biological race, and it still is. But it is far from unassailable, as we shall see.

> *The* Fst *statistic measures the degree of genetic differentiation between two given populations.*

Box 1: *F* Statistic

A method proposed by Wright (1951) to describe population structure with 3 *F* statistics: *F*is, *F*it and *F*st, whose relationships are established as follows:

$$(1 - F\text{it}) = (1 - F\text{st})\,(1 - F\text{is})$$

*F*is is the correlation between alleles* carried by homologous* chromosomes* among individuals in the local population, and *F*it is the correlation in the total population. These indices are also called fixation indices (Fi) and describe the deviations from Hardy-Weinberg equilibrium* predictions in local populations compared to the total population:

$FI = (h_o / h_e)$

With *ho* = observed heterozygosity and *he* = expected heterozygosity.

*F*i = 0 corresponds to Hardy-Weinberg equilibrium; if *F*i > 0, there is a deficit of heterozygotes, due either to inbreeding (*F*is > 0) or to inbreeding + population heterogeneity (*F*it > 0).

The *F*st component is a commonly used measure of **partitioning the population into distinct subpopulations** (if *F*st =0, *F*it = *F*is), and can be interpreted as the variance* of allele frequencies between populations.

For readers who would be put off by this very mathematical explanation, in common language, *F*st is a measure of the genetic differences between two populations, for example, between individuals of European and African ancestries. The higher the *F*st is, the more important the genetic differences are between the populations, statistically speaking. The number 0 indicates the absence of genetic differentiation, and 1 shows that the two populations differ totally in their genetic polymorphism.

Lewontin's Error (Sophism?)

"Lewontin's fallacy" is a qualifier used in a famous article by Anthony W. Edwards (2003). I introduce the slightly teasing translation of "sophism," because, given Richard Lewontin's immense competence, one cannot help but suspect that he knowingly oriented his interpretation with the goal of discrediting the concept of race.

> *Lewontin's error is not having taken into account an essential deficiency of the* F*st statistic: it fails to recognize the* *correlation* *between the genetic markers that differentiate geographical populations.*

Edwards's publication (2003) highlights a remarkable fact that does not appear in Lewontin's article. *F*st is what is called a "black box" in statistics. It does not shed light on certain remarkable properties of the sample under study. What Edwards emphasizes is that the genetic variants that account for the 5–15% differences between continental populations ("races") are *correlated*, congruent: their polymorphisms all move in the same direction. In other words: knowing the genetic profile of marker A allows one to anticipate with a high probability the profile of markers B, C, D, etc. This is the same phenomenon as knowing that a man has frizzy hair allows one to anticipate that he does not have milk-white skin. This property of genetic markers makes it possible to reliably characterize the major geographical populations—Europeans, Africans, Asians, etc. (see Figures 4–8).

Edwards's work (2003) was attacked by Lala and Feldman (2024), who considered that this author "confused geographical origin and race." However, this semantic quarrel does not detract from the argument of character congruence and its neglect by the *F*st statistic (Edwards, 2003).

The Correlation (Congruence) Between Genetic Markers Differentiating Geographical Populations Allows for the Establishment of "Genetic Genealogies" (Evolutionary Trees) Grouping them According to their Degree of Genetic Relatedness

This remarkable property of congruence between genetic markers, completely overlooked by the *F*st statistic, makes it possible to quantify the genetic differences between geographical populations, and these differences can be visualized by various diagramming processes: dendrograms, "phylogenetic trees." Some examples are given below.

An interesting fact to note: Figure 4 is inspired by pioneering work by Masatoshi Nei (1978), based on a limited number of genetic loci* (11 loci concerning proteins and 11 loci corresponding to blood groups), whose resolution is much lower than that of modern genomic markers. However, this crude marking clearly identifies the main human geographical populations. It is therefore false to say that these geographical groups can only be distinguished from one another by the use of an extensive array of the most modern molecular markers, as Jordan (2008) claimed.

We note in passing that Masatoshi Nei, a distinguished Japanese American population geneticist, unashamedly uses the term "human race." He will be far from the only one, as we shall see.

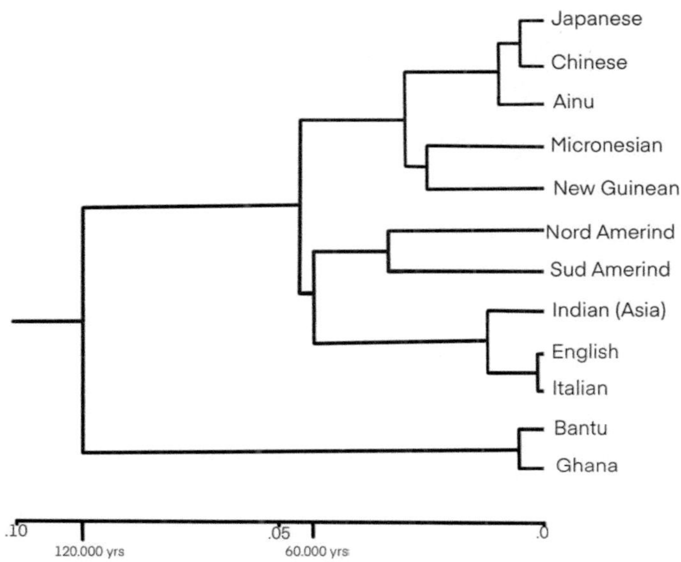

Figure 4: Phylogenetic tree (= genetic genealogy) of different geographical populations (Figure author: Jenny Telleria, IRD Montpellier, France; adapted from Nei, 1978).

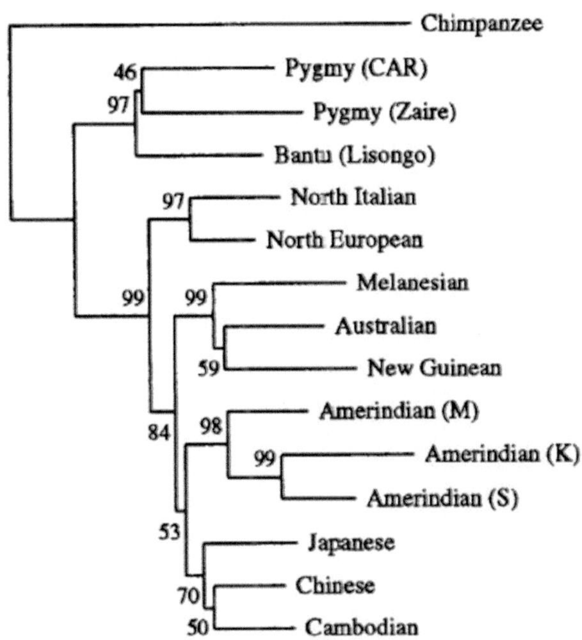

Figure 5: Genealogical tree of different human geographical populations characterized by a high number (123) of diversified molecular markers. The numbers indicate the robustness of the branches (Figure author: Dr. Jenny Telleria, IRD Montpellier, France; adapted from Nei and Takezaki, 1996).

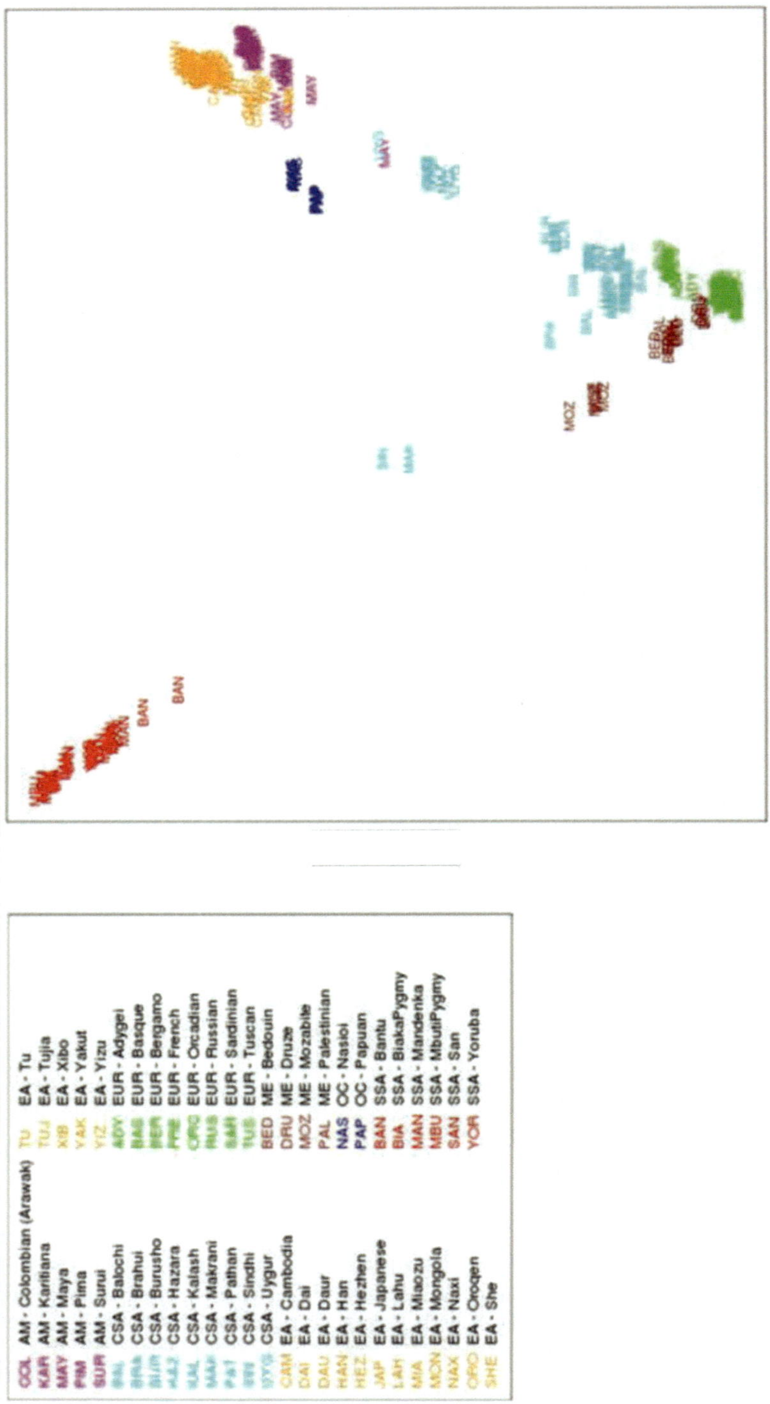

Figure 6: Visualization by "principal component analysis" (a statistical approach) of the genetic relationships between 255 individuals from 51 geographical populations characterized by 954,063 SNPs: America (AM (pink), Central and South Asia (CSA; light blue), East Asia (EA; orange), Europe (EUR; green), Middle East (ME; brown), Oceania (OC; dark blue), Sub-Saharan Africa (SSA; red). As in Figure 4, the grouping according to continental origin appears clearly (adapted from López Herráez et al., 2009).

M. Nei revisited the same question 18 years later (Nei and Takezaki, 1996), but this time based on a considerably larger sample of genetic markers, including markers tracing DNA* itself, which was allowed by the progress of molecular biology. The results fully align with those of 1978 (Figure 5). Here again, the authors naturally use the term races, and even "major races," in this case Europeans, Africans, and Asians. This is reminiscent of the "grand'races" (major races) of classical physical anthropology (Vallois, 1976).

The most modern molecular markers corroborate the results of the "ancestors" of the 1970s–80s in terms of visualizing the degrees of relatedness of geographical populations.

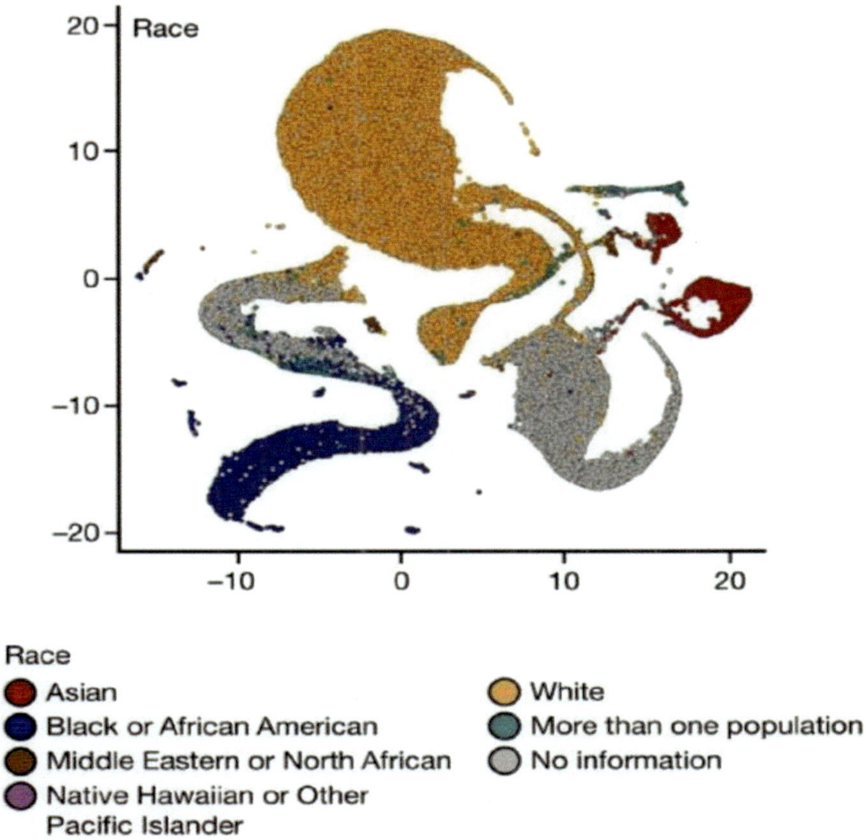

Figure 7: Characterization of the main human geographical populations by state-of-the-art high-resolution genomic typing. As in the case of the studies cited in Figures 4, 5 and 6, the geographical groups distinguished by physical anthropology are clearly visible. African Americans appear grouped with Africans ("black"). The authors use the term "race" to designate the geographical groups (Bick et al., 2024).

HLA* polymorphism (Human Leucocyte Antigen*) also corroborates the geographical subdivisions (Figure 8).

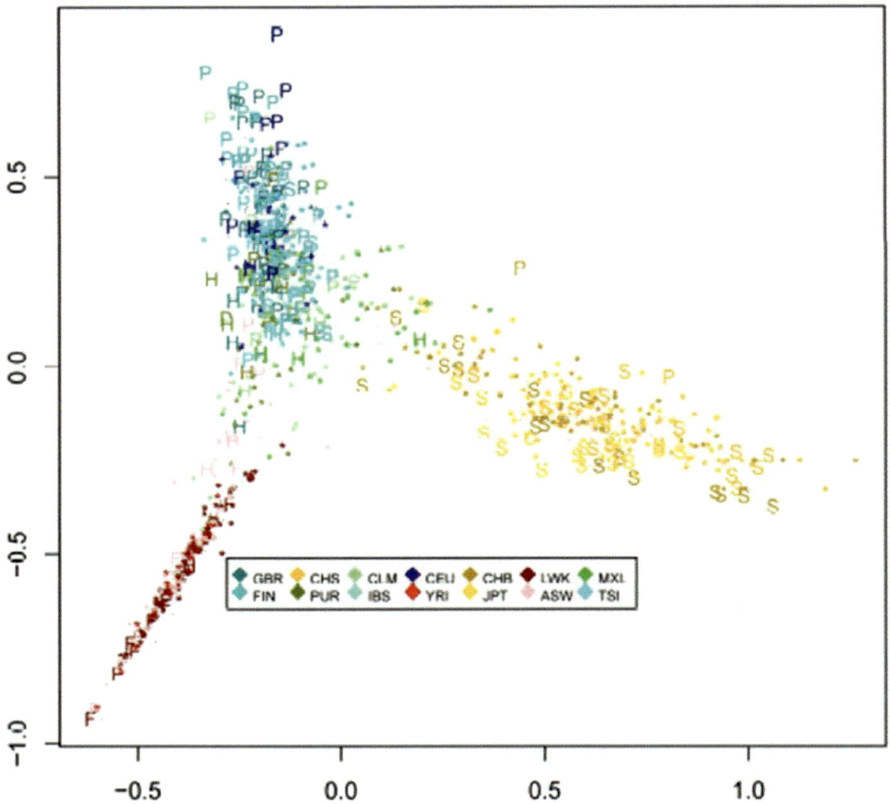

Figure 8: Principal component analysis of geographical populations by HLA (Human Leucocyte Antigen) polymorphism. Africans (bottom left), Asians (bottom right) and Europeans (top) are clearly distinguished. Adapted from Gourraud et al., 2014.

This ability of genetic markers to align with the "classic" geographical groups (Africans, Europeans, Asians, etc.) had already been noted by other studies (Rosenberg et al., 2002, 2005; Rosenberg, 2018, 2020). The precision of these identifications increases with the number of markers used, following the principle of "congruence" (Rosenberg, 2018).

However, this possibility of identifying geographical groups by genetic marking has been questioned by certain authors (Serre and Pääbo, 2004). The geneticist Joseph L. Graves postulated that the subdivisions highlighted by genetic markers depend on the initial choice (Anonymous, 2024a). If one chooses the main continental groups (Africans, Europeans, Asians), the genetic subdivisions will align with these continental groups. This argument was contradicted by Rosenberg et al. (2005). Raff (2014) criticized the work of Rosenberg et al. (2002) by arguing that the interpretation of the STRUCTURE software used by these authors was open to criticism, particularly for the number of units of analysis that this software individualized. But precisely, what is striking is the remarkable convergence observed between very diversified genetic markers, whether they are "outdated" like those used by Nei (1978) or, on the contrary, "state-of-the-art" (Bick et al., 2024), and between highly varied analysis treatments. All these studies show the same strong trend: an indisputable association between the genetic profile of populations and geographical origin.

> *If "races" had no genetic reality, there should be a total absence of correlation between genetic diversity and "races."*

Despite the reservations expressed by Serre and Pääbo (2004), Graves (Anonymous, 2024a) and Raff (2014), it is evident that the fact that genetic markers tend to reliably identify the main human geographical populations considerably diminishes the scope of Lewontin's argument. Indeed, if geographical populations ("races") had no genetic reality, there should be a total absence of association between historically identified geographical populations (Africans, Europeans, Asians) on the one hand, and genetic diversity on the other.

One must be careful not to make these visual representations of human genetic diversity say more than what they represent: they are merely genealogies, pedigrees, which indeed reflect the history of populations, their migrations, and the importance of genetic exchanges between them. They are visual summaries of all these processes. No general conclusions about the essential phenotypic properties of these populations can be drawn from them.

Is it Acceptable to "Tweak" Scientific Results "for a Good Cause"?

I am somewhat anticipating the second part of this book ("Ethics and Politics"). Lewontin was, as I have noted, an outstanding geneticist. It is reasonable to wonder if he deliberately concealed the remarkable property of congruence of genetic markers (which the *F*st statistic does not highlight) in his 1972 article. I am not claiming that he did, only that we cannot rule it out. Wasn't Lewontin described as "dogmatically non-dogmatic"? (Berry and Petrov, 2021). Might this not make Lewontin's fallacy a bit of Lewontin's sophism?

This brings us back to questions that will run throughout this book: Is it acceptable, "for a good cause," to "tinker" somewhat with factual data, possibly through deliberate or even unconscious omission, in the name of a morally estimable cause, in this case, antiracism? Of course, this question applies a fortiori if the cause is not estimable (such as supporting racism). More broadly: Are we inclined to believe a hypothesis more because what it advocates is morally defensible, and conversely, do we have the opposite tendency to reject a hypothesis because it seems to support unsympathetic facts? This would be a specific case of the "myside bias" (also called confirmation bias), a bias that pushes us to seek at all costs to confirm a hypothesis that flatters our convictions (Orlando, 2021). The entire question of the independence of science on the one hand, and morals and ethics on the other, is thus raised. But let us not anticipate.

> *Lewontin did not take the turn of the advent of genomics and remained a fierce opponent of any genetic determinism.*

Richard Lewontin always adhered to an argument primarily based on low-resolution markers. He little discussed the contribution of the advent of genomic markers. Yet, as we will see later, these markers completely reshuffled the deck for the *F*st argument. It is noteworthy that until a relatively recent date (anonymous, 2002b), Lewontin was fiercely opposed to the Human Genome Project (HGP) (Cavalli-Sforza, 2005), a project aimed at sequencing the entire human genome. Lewontin accused the HGP of resting on a presumption of strict biological determinism (the belief that all properties of the human being, including cognitive traits, are 100% determined by genes). This excessive stance, this absolute rejection of any innate determinism, led this highly valued scientist to make resolutely false assertions, such as claiming that skin cancers only appear after reproductive age and therefore cannot be an agent of natural selection* (anonymous, 2002b). However, melanoma, the most malignant of all tumors, is the most frequent cancer in young adults between 25 and 50. This false hypothesis led him to deny that natural selection, through climatic adaptation (degree of sun exposure), could explain differences in skin color between geographic groups. He attributed these differences to sexual selection*. Yet, the hypothesis of climatic adaptation for skin color is very generally accepted (Takezawa, 2023), which does not exclude a possible influence of sexual selection, also invoked by Harpending (2002).

Another Bias of the Fst *Statistic: Its Extreme Versatility*

To reiterate, the "*F*st argument" can be summarized as follows: geographical differences *within* geographical populations ("races" in the current debate), measured by the *F*st statistic, are much greater than those observed *between* these geographical populations, which would invalidate the concept of biological race. However, another feature of this statistic further weakens this argument: it is extremely sensitive to local demographic conditions. For instance, Moreno-Estrada et al. (2014) observe that between two Mexican Amerindian populations, the Seri and the Lacandons, the *F*st is 0.136, whereas it is only 0.11 between two major geographical populations, the Europeans and the Chinese. Yet, it would not occur to anyone to consider that there are more biological differences between Seri and Lacandon than between Europeans and Chinese. This surprising result concerning these two Native American populations is undoubtedly attributable to specific local demographic factors. Nevertheless, this example clearly shows that the *F*st statistic must be interpreted with caution.

Other Champions of Scientific Antiracism: Franz Boas, Albert Jacquard, Jacques Ruffié

The defenders of scientific antiracism prove to be eminent scholars.

The least one can say is that this camp, an undoubted defender of a noble cause, recruits "heavy hitters." The scientists cited, despite their (numerous) detractors, are among the most respectable of their time.

Presenting these personalities will allow us to lay out the scientific arguments they advanced to attempt to disqualify the concept of a biological human race.

- Franz Boas (1858–1942), pioneer of scientific antiracism in the United States between the two World Wars. Boas called racism the "witchcraft of our time" (Murray, 2020). It is therefore not surprising that he was a favorite target of "race realists" (Sarich and Miele, 2004; Wade, 2014). Of German origin, a physicist by training, he is considered one of the founding fathers of American cultural anthropology. He was one of Claude Lévi-Strauss's inspirations. His convictions are characterized, as is the case for Richard Lewontin, by a rejection of all biological determinism, and also by the "indivisible nature of humanity" according to the historian Charles King (Bloch-Lainé, 2022). Following a study trip among the Inuit, he was convinced of the "relativity of all education." A man of the field, he led numerous study trips among different peoples, which persuaded him of the impossibility of hierarchizing different civilizations: each people faces the challenges of life with their specific cultural responses. None of these responses is superior to others in the absolute. I tend to share this view, after my long and repeated contact with diverse cultures. In this regard, according to a personal experience of mine, the Native Americans of French Guiana, at least those who continue to live a hunter-gatherer lifestyle, are undoubtedly not very versed in Kantian philosophy or integral calculus. On the other hand, due to the necessity of having an intimate knowledge of the surrounding environment on which their survival depends, they have developed a very sophisticated binomial nomenclature, reminiscent of Linnaean taxonomy, to identify all the living beings around them. An average French person plunged into the same environment would be extraordinarily defenseless there.

Boas's Hypotheses on the Non-Heritability of Skull Shape Have Been Questioned*

Tasked by the U.S. Congress in 1911 to study immigrants, Boas undertook to prove using physical anthropology techniques (measurements of bone dimensions, the skull in particular) that many physical traits were modulated more by the environment than by heredity (Bloch-Lainé, 2022). However, this "proof" was strongly questioned by a recent study (Sparks and Jantz, 2002), which, by re-examining Boas's own data with modern statistics, arrived at exactly opposite conclusions, finding that the variability of cranial morphology appeared to be highly heritable, meaning partly modulated by genetics (see the concept of heritability in the glossary). These later results were discussed, but not falsified, by Holloway (2002).

Furthermore, Boas considered that differences in behavior were 100% cultural (Wade, 2014).

Franz Boas had a considerable influence on the ideas of his time. He had numerous disciples, including two very well-known ones, Margaret Mead and Ruth Benedict.

> *Ruth Benedict, a student of Boas and thus part of the pre-war antiracist movement, did not equate the study of races with racism.*

I recall that the latter, though a determined antiracist, did not consider the study of races to be a racist approach in itself, whereas since the 1970s, the mere act of speaking of race is often considered racist (Taguieff, 2018).

- Albert Jacquard (1925–2013) is a French Population Geneticist. His detractors, particularly those from the French "New Right" movement, have repeatedly tried to challenge him by questioning his scientific competence, entirely in vain. According to my American evolutionist colleagues, Jacquard, as an evolutionary geneticist, was among the greatest, his reputation being underserved by a most honorable modesty and discretion.

Jacquard, in addition to being an outstanding scientist, was a true humanist, entirely devoted to numerous causes, particularly the homeless. He was a highly estimable man, without a doubt. I would have been very honored to debate with him. His most notable social struggle was in the service of "scientific antiracism," of which he was a leading Figure.

> *Jacquard sought to disqualify the concept of race through the discordance of traits: for example, a classification based on skin color did not coincide with another based on blood groups.*

While Albert Jacquard's humanism is highly laudable, his scientific approach to antiracism is vulnerable to criticism. It is based above all on the argument of the *discordance of traits*: if one takes a certain trait, for example a certain blood group, one will obtain a certain grouping of geographical populations. If one takes another trait, for example skin color, one will observe a different grouping. If one takes a third trait, for example lactose tolerance (the ability to digest lactose, the sugar in milk), one will have a third grouping. This would lead to an infinity of racial groupings, which is supposed to demonstrate the futility of the concept of race (Jacquard, 1978, 1982). This argument, clever at first glance, can only mislead the layman. A Master biology student who would advance it to discuss newt or beetle subspecies would get a failing grade. Discordance is verified if one selects certain specific cases. But such a discordance of traits can even be frequently observed between *different species*, which considerably weakens the use of the argument to disqualify the concept of biological race. Furthermore, this discordance is globally, *statistically* invalidated if one considers traits not in isolation, but as a whole. They tend to be *correlated*, as we have seen concerning genetic markers. Moreover, even if one takes a single set of traits, such as blood groups, they roughly cover the major geographical groupings identified by physical anthropology (Boyd's classification, 1950, cited by Dobzhansky, 1962a). On the other hand, if one considers, for example, the blood group B alone, its distribution does not align with that of skin color (Anonymous, 2024a). Subspecies, or geographical races, should not be described by *isolated traits*, but by *sets of traits* (Ayala and Kiger, 1984).

According to Jacquard, it is urgent to eliminate the concept of race, an instrument that is not only useless, but *dangerous* (Taguieff, 1988). This argument, far more convincing than the discordance of traits, cannot be dismissed with a wave of the hand.

Professor Jacquard, an eminent scientist on the one hand and a highly estimable humanist on the other, may have allowed these two aspects of his engaging personality to somewhat osmose.

- Jacques Ruffié (1921–2004) Professor Ruffié was a French hematologist and geneticist. He is the founder of hemotypology, a discipline that consists of characterizing human populations by blood markers. Like Albert Jacquard, he is one of the most ardent French defenders of "scientific antiracism" (Taguieff, 1988). All data below are taken from his 1983 work: De la biologie à la culture (From Biology to Culture).

Ruffié takes up the argument of trait discordance, defended by Jacquard, with an inadequate example.

The author cites a study by Italo-American anthropologist Lluigi Cavalli-Sforza, showing two human population trees. One of the trees is based on the quantification of 58 blood genetic markers, the other is based on 28 external morphological measurements. Ruffié uses the argument of a non-total fit between the two trees to reject the concept of human races. However, the two trees are remarkably similar. They are definitely not independent of each other. The groups they individualize are the same. One does not expect 100% concordance between two analyses based on traits that rely on such different genetic mechanisms and have undergone totally dissimilar natural selection pressures.

Ruffié uses another argument to disqualify the concept of human races: according to him, human genetic variability is a perfect continuum. There are therefore no strict boundaries between "races." This is the "typologist" argument, which will be developed later (see: "Other Arguments of Scientific Antiracism").

Ruffié was a fierce opponent of all biological determinism.

Like many other defenders of scientific antiracism, this scientist was also fiercely opposed to any idea of biological determinism. This led him to defend fundamentally inaccurate ideas, for example, that the adaptation to altitude of certain populations, the Native Americans of Peru and Bolivia and the Tibetans, was "modest," and based on non-genetic acquired characteristics. This is quite simply false. Another proposition to be rejected: the abundant melanin in black skin would, according to Ruffié, be a "derisory" protection compared to appropriate clothing. This is contradicted by the fact that the incidence of malignant melanomas (an extremely aggressive skin cancer), according to data from the Centers for Disease Control (CDC) in Atlanta, is much higher in Americans of European descent than in any other geographical group (https://search.cdc.gov/search/?query=melanoma&dpage=1).

In the same vein, Ruffié defends the idea that cultural evolution has completely supplanted biological evolution in humans, and that humans are now exempt from the pressure of natural selection. This is a very widespread idea, even among eminent scientists like the paleontologist Stephen J. Gould. It is highly debatable, as we shall see later (see: "Other Arguments of Scientific Antiracism").

Contrary to a widespread idea, which holds that racism is a recent invention to support colonialism, the author believes that interethnic racism dates back to antiquity.

> *Ruffié predicted a generalized mixing of human populations ("white coffee process").*

According to Ruffié, who speaks of the "difficult end of a myth" (that of human races), the human species is in the process of "deraciation," with the different geographical populations tending to merge. He uses the picturesque term "caféaulisation" (turning to coffee-with-milk color) for this, and envisions a not necessarily exciting dystopian future, in which the whole of humanity will have become homogenized into a uniform type, swarthy, with subtly slanted eyes. Good heavens! No more Cate Blanchett, no more Omar Sy, no more Bruce Lee? What a chilling nightmare! However, the author, magnanimous, does not go so far as to advocate universal compulsory miscegenation. No reverse apartheid. Phew!

Phenotypic Expression is Not a Gene Election by Universal Suffrage

The *F*st argument is based on the hypothesis that the "weak" genetic differences observed between major geographical populations imply that there are, all things considered, few general differences ("negligible" differences) between these populations. However, as we have seen, the convergence between the genetic markers that differentiate geographic populations (the 5–15%), not highlighted by the *F*st statistic, weakens this argument. Furthermore, it is not at all evident that these "weak" differences reflect the overall characteristics of the studied populations: the differences between these populations have been described, not by focusing on their *F*st, but rather, on their phenotypic differences. However, *phenotypic expression is not a gene election by universal suffrage*. Minimal genetic differences can lead to considerable phenotypic differences. Furthermore, the analysis of other genetic and biological characteristics (low-frequency* and rare variants*, structural variation*, gene regulation* and epigenetics*; see below) does not really corroborate the "weakness" of the *F*st differences discussed above.

It is important to emphasize that human geographic populations, whatever names are given to them, have never been described on the basis of any percentages of genes, but on their *genetically heritable phenotypes*.

The Contribution of Low-Frequency and Rare Genetic Variants

The "slam-dunk argument" of scientific antiracism to disqualify the biological concept of race is the famous *F*st statistic (see above: "The apportionment of human diversity: the *F*st argument," and Box 1). In short: the greatest part of human genetic variability is found, not *between* geographical populations, but *within* each of them. In the context of this book: there may be more genetic differences between some "whites" than between some "whites" and some "blacks." In this perspective, the risk is not zero of finding more kinship ties between President Donald Trump and Chinese President Xi Jinping or African American actor Eddie Murphy than between the same Donald Trump and one of his supporters wearing a red "Make America Great Again!" cap. Bad news for white supremacism.

This argument was mocked by Sarich and Miele (2004), who said that with such a statistic, one could find more genetic kinship between a chihuahua and a Great Dane than between two chihuahuas.

We have already seen that the *F*st argument was weakened by the fact that the genetic variants differing between geographical populations are *correlated*, meaning they all go in the same direction, a particularity that the *F*st statistic overlooks. It is this property that allows the principal human geographical groups to be reliably characterized through "genetic genealogies" (Figures 4–8).

> *The* Fst *"slam-dunk argument" is weakened by the role played by low-frequency and rare variants.*

Another major bias of this *F*st is that, particularly in the foundational study by Lewontin (1972), it is based on so-called "common" variants, meaning those found in all geographical populations, and whose general frequency of their least frequent allele is greater than 2%. Many of these common variants* are ancient, dating back more than 100,000 years, thus predating the "out-of-Africa" (Gurdasani et al., 2019; see further "out-of-Africa"). These definitions, based on their frequencies, of different categories of variants may differ slightly from one author to another. These "common" variants are also characterized by a low resolving power (Figure 3). In the pioneering era in which Lewontin wrote his famous article, geneticists only had these to work with. These variants, besides being "common" and of low resolution, have another particularity. They are generally "selectively neutral," meaning they are not subject to natural selection, and are "probably of little importance for phenotypic differentiation" (Lala and Feldman, 2024). This is particularly true for isozymes, which were the "workhorse" of population geneticists from the 1970s onwards. These markers have no direct influence on the phenotype (all the essential properties of organisms, such as height or skin color). They are a kind of genetic "tick-tock" beating at a fairly regular rhythm. Because of this, they can be considered "temporal" or "historical" markers. They are very effective for reconstructing the history of populations and their reciprocal filiations, to establish genetic genealogies.

Unlike isozymes, microsatellites, another category of "common," "historical" markers, can play a role in phenotypic expression, particularly in various pathologies (Xie, 2024). Like isozymes, but with a higher resolving power than the latter, they are excellent population markers. They differentiate Europeans, East and South Asians, Native Americans, and Africans into different groups or "clusters*" (Xie, 2024; Shi et al., 2023). They are also widely used in forensic medicine, for the identification of criminals (Budowle and Sajantila, 2024) or for paternity testing.

The immense progress in genomics has unearthed the existence of other categories of variants that are of considerable evolutionary interest: these are "low-frequency" variants (frequency between 0.5 and 2%), and "rare" variants (frequency less than 0.5%) (Nelson et al., 2012; Tennessen et al., 2012). These newcomers present two characteristics that are considerably informative for the present discussion. First, unlike "historical" variants such as isozymes, they have a strong impact on the phenotype, including on "complex" diseases*. Thus, these rare variants are implicated in height (Marouli et al., 2017), susceptibility to certain drugs (Manry and Quintana-Murci, 2013), the genesis of diseases such as Alzheimer's, Parkinson's, psoriasis, autism, schizophrenia, as well as in susceptibility to AIDS and other infectious diseases (Gonzaga-Jauregui et al., 2012; Manry and Quintana-Murci, 2013). Second, these markers, defined by their frequencies at the general population level, tend to be specific to given geographical populations, in other words, to show strong frequency differences between populations (González Burchard et al., 2003; Hindorff et al., 2018; Manry and Quintana-Murci, 2013; Martin et al., 2017; Mersha and Abebe, 2015; Nelson et al., 2012; Sudmant et al., 2015; Tennessen et al., 2012). Up to 86% of these variants are found in only one continental group (The 1000 Genomes Project Consortium, 2015).

Consequently, the *F*st statistic, the emblem of scientific antiracism, will take on a completely different appearance if it is based on these markers. This cannot be swept under the rug given their considerable evolutionary importance: if we take these markers into account, we will see far more genetic differences *between* populations than *within* each of them, in other words: an image exactly inverted compared to that of Lewontin's historical "paper" (1972). It should be noted that these rare and low-frequency variants constitute the largest part of the genetic variants known in the human species (Abdellaoui et al., 2023).

The Impact of Structural Variation

A long-underestimated player in human variability

Structural variation deals with long fragments (≥ 50 base pairs) of the DNA sequence. It is still imperfectly known.*

The so-called structural variation does not concern point mutations*, i.e., the variation observed at individual nucleotides in a DNA sequence, but rather the variability concerning long segments of DNA, whose length exceeds 50 base pairs and can exceed 3 megabases (one megabase = one million bases) (Conrad and Hurles, 2007). Structural variation thus differs from variation affecting a single nucleotide and short insertions/deletions (indels) (less than 50 base pairs) (Collins and Talkowski, 2025). This structural variation is still imperfectly understood. There are two main reasons for this: (i) as in all genomic research, sampling remains highly unbalanced, with an overrepresentation of individuals of European ancestry (see below: The "genetic discovery bias"; too many "whites"?); (ii) for technological reasons, the identification of structural variations is difficult in the 10% most repetitive parts of the genome (Collins and Talkowski, 2025).

Structural variation includes inversions*, duplications*, and insertions/deletions (indels)*. Duplications and indels, referred to as copy number variation* (CNV*), result in changes in the total quantity of DNA, which can have significant phenotypic consequences. In contrast, inversions do not change the overall DNA dosage. They only change the orientation of the concerned DNA sequence.

Structural variation seems to play an important role in phenotypic expression and to be subject to natural selection.

These phenomena are much more common in human populations than previously thought. Like low-frequency and rare variants, they seem to have a stronger impact on the phenotype than variation at individual nucleotides (Vasseur and Quintana-Murci, 2013), especially since this structural variation appears to be strongly subject to natural selection (Collins and Talkowski, 2025). This type of variation does not only concern pathological traits and is widespread in healthy individuals. If one considers the totality of variation at individual bases, it is more affected by the individual bases carried by the structural variation than by the variation at individual bases occurring independently of this structural variation (Collins and Talkowski, 2025; Sudmant et al., 2015). Lastly, the geographical distribution of structural variation parallels that of variation at individual bases: differences in frequencies are observed from one geographical group to another, and the rarer the structural variants, the more specific they are to populations (Sudmant et al., 2015). This latter phenomenon can only further weaken the "knockout argument" of *F*st (Lewontin, 1972).

The Role of Gene Regulation and Epigenetics in Differentiation Between Geographical Populations

Gene Regulation: A Major Factor in the Genesis of Phenotypes

Apart from the impact of rare and low-frequency variants and that of structural variation, another phenomenon is likely to weaken the "*F*st knockout argument." We have seen that the "common" or "historical" genetic variants on which this *F*st statistic was based, particularly isoenzymes, play a weak or non-existent role in phenotypic differentiation.

Gene regulation and epigenetics modulate the expression of phenotypes without modification of the DNA sequence of the genes concerned.

In contrast, gene regulation and epigenetics, which condition the level of gene expression, seem to have a considerable impact on phenotypic differences between populations and species. Gene regulation is the phenomenon that explains why, in a given individual, cells from different organs, for example, those of the liver and the heart, are different, even though they house the same genetic equipment. Gene regulation is thus at the origin of the differences between liver cells (hepatocytes) and myocardial (heart) cells. This was revealed by the historical work of Jacob, Lwoff, and Monod (Nobel Prize in Medicine in 1965) on gene regulation in bacteria (the "Operon system"). Depending on the quantity of nutrients (sugars) available in the culture medium, *Escherichia coli* (the "colon bacillus"), the bacterium they used in their experiments, "turns off" or "turns on" the gene that synthesizes the enzyme* β-galactosidase, which is required for sugar metabolism. This "biological switch" is under the control of the famous "regulator genes." Regulator gene + coding gene* = operon. As the charismatic Nobel laureate Jacques Monod liked to say, "what is true for *Escherichia coli* is true for the elephant." This aphorism is verified for gene regulation, which plays a major role, not only in microbes but also in so-called "higher" organisms, including humans. For example, it is gene regulation that intervenes to allow the caterpillar to become a butterfly (Henry, 2019).

Phenotypic expression results from a constant interaction between genotype and environment

In *Homo sapiens*, as in any living being, phenotypic expression results from a constant interplay between the genotype and the environment ("G x E"), particularly through the mechanism of gene regulation (Boye et al., 2024). For example, sunlight (an environmental parameter) allows one to acquire a "tan." But this ability is unequally distributed according to the genetic makeup of each individual, and this is largely related to geographical origin. The author of this book would tend to turn an incandescent copper color, while his Siamese wife turns deep brown in only a few minutes when exposed to the sun.

In their historical article, King and Wilson (1975) estimated the percentage of nucleotide differences (a reliable measure of "genetic distance*") between humans and their closest relative, the chimpanzee, at 1.1%. In other words, from the perspective of DNA sequences, humans and chimpanzees are 98.9% similar. This measure appears to be reliable, as recent genomic studies give a very similar order of magnitude (Chimpanzee Sequencing and Analysis Consortium, 2005; Tung, 2024). King and Wilson (1975) attributed a major part of the considerable phenotypic disparity observed between humans and chimpanzees to differences in gene regulation. Fu and Akey (2013) indeed postulated that the differences in gene expression between humans and chimpanzees were considerable. This would be particularly true for brain genes.

Similarly, a significant discrepancy is noted between the low *F*st observed between human geographical populations and their striking phenotypic differences. Sarich and Mile (2004) noted that there were more phenotypic differences between Africans and Asians than between chimpanzees and bonobos, which are different species. Harpending (2002) observed that an *F*st adapted to measure differences in skin color was 0.6 to 0.9. It is highly probable that gene regulation phenomena play an important role in the phenotypic differences between human geographical groups.

Structural genomic variation seems to play a large role in gene regulation (Collins and Talkowski, 2025; cf. "The impact of structural variation").

The significant action of gene regulation would apply equally to healthy subjects and to pathological traits. In evolution, "soft sweeps*" concern natural selection acting on pre-existing genetic diversity ("standing variation*"), on the action of numerous genes with limited individual effect, and on small, regulatory, polygenic*, epistatic*, or epigenetic changes. It appears that this phenomenon has acted more on human evolution than "hard sweeps*", which, in contrast, denote the sudden appearance of new mutations with a drastic adaptive effect (Fu and Akey, 2013). The phenomenon of "soft sweeps" is, for example, the one in action in the selection of domestic animal breeds. The breeder does no more than artificially accelerate a natural phenomenon by selecting, in each generation, the subjects possessing the phenotypic qualities they wish to develop.

To illustrate the impact of gene expression on the geographical differentiation of human populations, there are highly significant differences between Asians and Europeans for gene expression in lymphoblastoid cell lines (Spielman et al., 2007). Lymphoblasts are precursors of lymphocytes, a type of white blood cell. They can be "transformed," i.e., made cancerous, for certain experiments. Benjamin et al. (2024) compared the gene expression of brain cells between Americans of European and African descent on post-mortem samples. The genes expressed differently between the two groups are involved in immunological defenses and in the genesis of vascular tissues. Garcia-Pérez et al. (2023) also found significant differences in gene regulation, not only between geographical groups (Americans of European and African ancestry) but also according to sex and age. Highland et al. (2022) obtained similar results by taking as study units Europeans, African Americans, "Latinos" (a heterogeneous group including various populations of Latin American descent), Asians, and Hawaiians.

Epigenetics: Lasting and Potentially Hereditary Modifications of Phenotypic Expression

Epigenetics concerns phenomena of gene activation and deactivation without any change in the corresponding DNA sequence. Epigenetic changes act more long-term than classical genetic regulation. They are sometimes transmissible from one generation to the next, which goes against the classic belief that only DNA changes are heritable.

> *The lifestyle of the parents, including the father, or even the grandparents, can influence the phenotypic expression of the offspring through epigenetics.*

In other words, the parents' experiences, recorded during their lives in the form of epigenetic imprints, can be transmitted to subsequent generations. Thus, the lifestyle of the parents, including the father, or even the grandparents (smoking, alcoholism, diet, overweight) can have an influence on the development of their offspring (Rambert, 2022). Meloni (2017) uses the term "soft heredity" to refer to epigenetics. Étienne Danchin (Paul-Sabatier University of Toulouse, France) proposes the concept of "inclusive heredity," integrating genetics, epigenetics, and cultural evolution (dos Santos et al., 2021).

This influence of the environment through epigenetics evokes a resurgence of the 19th-century theory of degeneration, which insisted on the harmful influence of exposure to unfavorable environments (the French philosopher of science Claude-Olivier Doron, cited by Mary, 2018a).

For Quintana-Murci (2021), there are innate differences in epigenetic processes between geographical populations.

Bantignies and Giacomo (2025) nonetheless emphasize that the hereditary transmission of epigenetic traits in humans has not yet had a conclusive demonstration at the molecular level. These authors recommend continuing research in this area using animal experimental models.

Non-genetic Phenotypic Differences Between Geographical Populations?

Considering the epigenetic hypothesis, Duncan et al. (2024) hypothesized that the adverse environments (chronic stress, deplorable economic conditions, starving) in which enslaved Africans lived could have epigenetic repercussions on their current descendants, which would explain certain phenotypic specificities of African Americans, without a genetic basis. This hypothesis was also defended by Meloni (2017), for example, to explain the lower birth weight of African Americans.

Raff (2014) inferred that the stress caused by systemic racism could promote high blood pressure in the populations who are victims of it.

> *The stress caused by slavery and systemic racism could have an impact on the phenotypic expression of African Americans through epigenetics.*

Other authors insist on the potentially transmissible epigenetic modifications that the still-current "systemic racism" and its deleterious effects on physical and mental health could cause (Pontarotti, 2021, 2023; Wright et al., 2022). Brown et al. (2025) claim that the differences between geographical populations regarding diseases or school success are attributable to this systemic racism.

These hypotheses are not implausible and need to be tested, for example, by analyzing the traits suspected of being the results of this slavery-related stress in biologically comparable African populations who did not undergo the slave trade.

Other Arguments of Scientific Antiracism

Scientists who seek to demonstrate that "human races do not exist" in the hope of emptying racism of its substance constantly use the same arguments.

No consensus among biologists on the non-existence of human races, despite what is often claimed.

In this debate, it is useful to recall, as already stated in the introduction, that, contrary to what is often heard, there is no consensus within the scientific community regarding the existence or non-existence of human races (Štrkalj, 2007). Leading scientists, such as Theodosius Dobzhansky or Ernst Mayr, who are not suspected of racism, have always defended the existence of human races. The statement: "Race and ethnicity are social constructs without scientific or biological meaning," a mantra of scientific antiracism and the official position of the *American Journal of Medical Association* (Flanagin et al., 2021) was denounced as false by Coyne and Maroja (2023). This stance is all the more remarkable given that the first author, Jerry A. Coyne, is a talented evolutionist, a student of the leader of scientific antiracism, Richard Lewontin. Coyne is known for his stand against apartheid in South Africa. He is unlikely to be suspected of racist thought.

Numerous biomedical publications continue to refer to the term and concept of race explicitly and specifically (see, for example, Bick et al., 2024; Huston and Raymond, 2013; a complete list of references at:
(https://drive.google.com/file/d/1qWNbPeyBCwDL9CgzoOMrHguFm9MeX4f2/view).

Even more tellingly, the hypothesis of the non-existence of biological human races is mainly prevalent in Western Europe and North America (much more so among social science specialists than among biologists), and much less so in Eastern Europe, Russia, and China (Štrkalj, 2007). It would be tendentious, even racist, to consider the scientists of these countries as of lesser value.

What can be noted, however, is the gradual abandonment of the term "race" used in isolation in biomedical literature (Byeon et al., 2021). The cited study only focuses on the journal *American Journal of Human Genetics*. From 1949 to 2018, the isolated use of "race" decreased from 26% to 4%, in favor of terms such as "African, Asian, European" which correspond to the formerly described "grand'races" (Vallois, 1976), or "race/ethnicity/ancestry (REA)" (Ali-Khan et al., 2011; Popejoy et al., 2018). This triple name suggests that each of its components is considered more or less synonymous.

One may doubt that the systematic search for semantic euphemisms is of any effectiveness in combating racism.

The reader can find: (1) a list of recent biomedical articles explicitly referring to the notion of race using the term itself:
(https://drive.google.com/file/d/1qWNbPeyBCwDL9CgzoOMrHguFm9MeX4f2/view);
(2) a list of articles which, without using the term race, replacing it with "ancestry groups" or "geographical populations," are based on sampling "major geographical groups": Africans, Europeans, Asians, etc.
(https://drive.google.com/file/d/1pffskqiM3WuPeD7gLIWYsSiXyTdwoeso/view).

In terms of biology and evolution, the only difference between these two lists is semantic. Both lists are far too long to be included in the bibliography of the present work.

Targeting the Wrong Target (Straw Man Strategy?)

To disqualify the concept of human races, "scientific antiracism" frequently attacks definitions that have never been used, either in physical anthropology or in population genetics. This is the result of either ignorance or a deliberate strategy.

> *Erroneous definitions of the concept of race put forward to discredit it: typological, phylogenetic, sociological concepts: the "straw man" strategy.*
>
> *The definitions of human races presented by physical anthropology and population genetics differ completely from the typological, phylogenetic, and sociological concepts, and are ignored by the argument of "scientific antiracism."*

The argument of scientific antiracism puts forward definitions of human races that are not those given by classical physical anthropology (Coon, 1962; Vallois, 1976) and by population genetics (Ayala and Kiger, 1984; Dobzhansky, 1962a; Hartl, 1985; King and Stanfield, 1990; King et al., 2012; Mayr, 1963, 2002), these definitions being systematically ignored in this argumentation. By doing so, the authors have no trouble showing that these definitions are not in agreement with the data, and thus intend to disqualify the concept of human race. One may suspect a typical "straw man strategy" here, whether unconscious omr concerted.

The Typological Concept of Human Races

This argument is repeatedly used by scientific antiracism: human races, according to this concept, are defined as entities, each of which is genetically quite homogeneous, and which are separated from each other by strict boundaries. This refers to the typological concept of the species, according to which a "type" (holotype) is described for each species, serving for the description of the species concerned, the type being deposited in a reference museum so that specialists worldwide can refer to it. Around the type, variability is considered negligible, a kind of background noise that must be disregarded. As noted by Taguieff (2018), the numerous authors attacking the concept of race based on this argument have no trouble demonstrating that human geographical populations in no way correspond to this picture (Anonymous, 2022a, 2024a; Brown et al., 2025; Combis, 2018; Donovan et al., 2024; Heyer and Reynaud-Paligot, 2019; Kierkegaard, 2019; Last, 2012; Lewis et al., 2022; National Academies of Sciences, Engineering, and Medicine, 2024; Pontarotti, 2021; Quintana-Murci, 2021; Raff, 2014; Ruffié, 1983; Sablonnière, 2020; Sauer, 1992; Stoppa-Lyonnet, 2015; Takezawa, 2023; Templeton, 2013; Wang et al., 2024; Wojcik, 2025; Wynshaw-Boris, 2020; Zou et al., 2023). The argument of the National Academies of Sciences, Engineering, and Medicine (2024), aimed at disqualifying the biological concept of race, is illustrative of this strategy: "Human genetic variation is continuous and overlapping, thus refuting the notion of *discrete* human races." (My emphasis)

This typological concept of human races has been accused of "racial Platonism" (Winegard et al., 2020), in the sense that it considers that a race should be endowed with an ideal "essence" in the Platonician sense of the term ("essentialism").

Such a typological definition of human races has never been used, to my knowledge, in population genetics. The authors belonging to this discipline who discuss the concept of biological race emphasize that within each race, genetic variability is high, and that there are no sharp boundaries between them (Ayala and Kiger, 1984; Dobzhansky, 1962a; Hartl, 1985; King and Stanfield, 1990; King et al., 2012; Mayr, 1963, 2002). Similarly, the anthropologist Henri Vallois (1976), in his classic work "Les races humaines" (Human Races) (Presses Universitaire de France), insists that each racial subdivision he identified shows great polymorphism, and that there are many intermediate populations between the main racial categories. Even the physical anthropology of the early 20th century, which was undeniably racist, noted that "races" showed great variability (Joubin and Robin, 1923). Faced with this typologist argument, one could also speak of a "decoy strategy," by analogy with those intercontinental missiles which, upon re-entry into the atmosphere, split their payload between the real bomb and multiple false bombs, the decoys, whose goal is to overwhelm defense radars.

This attack on the concept of race, relying in a concerted or unconscious manner on the "straw man/decoy" strategy, and based on an erroneous typological definition, turning its back on the definitions of population genetics and physical anthropology, appears to me to be a major bias of "scientific antiracism". However, this approach has had a considerable impact, and continues to greatly influence the debates.

The Phylogenetic Concept of Human Races*

This argument is primarily employed by Templeton (2012, 2013, 2017). According to this author, who I must specify is a first-rate scientist, human geographical populations, to deserve the term race, should meet two criteria: (i) there should be a level of genetic divergence of at least 25% between them; (ii) they should correspond to phylogenetic (evolutionary) lineages strictly separated from each other, which refers to the typologist argument. Of course, the author has no trouble demonstrating that human geographical populations satisfy neither of these two criteria. As we will see in the paragraph below: "The concept of subspecies or geographical race," the concept of geographical race has never been based on such requirements. Even going up to the taxonomic level above it, namely the species concept, it is not based on the requirement of any threshold of evolutionary divergence. Some species can be evolutionarily very close, and not satisfy the criterion of 25% genetic difference between them, if the phenomenon of speciation is very recent.

The major human geographical groups do not meet the very demanding criteria of Templeton (2012, 2013, 2017). Nevertheless, they are characterized by a very clear "phylogenetic signal*", as evidenced by the multiple and convergent constructions of evolutionary trees (Figures 4–8). These discontinuities indicate that the populations identified by these trees have been genetically isolated from each other by geographical distance for long periods of time (space/time isolation), even if this isolation has been only partial. This separation has led to genetic stratification, which must be taken into account in all applied studies.

A Sociological Definition of Human Races Partially or Totally Disconnected from Biology

This is an even more confirmed case of the "straw man strategy" evoked earlier. In everyday language, as we have seen, the word "race" has multiple meanings, as is the case for many complex concepts. In this profuse semantic range, it is easy to extract a meaning that is almost totally disconnected from any biological reference, and in any case, which has never been used, either in physical anthropology or in population genetics, and which refers much more to a concept of race considered from a sociological point of view. Moreover, since the origin of the debate, the concept of race has constantly oscillated between *sociological* and *biological* meanings (Pontarotti, 2023). This is an opportunity to recall here that the present book focuses on the *biological* concept of human races.

The work by Peterson et al. (2019), otherwise excellent, provides a definition of race that deals very much with the social sciences and very little with genetics. I quote in extenso: "Race: A culturally and politically charged term, *for which definitions and meaning are context-specific* (my emphasis: one could not say better). Race is related to individual and/or group identity and is often linked to stereotypes of visible physical attributes such as skin and hair pigmentation. The concept of race is tightly linked to social power dynamics and has historically been used to justify hierarchies of power, discrimination, and oppression in an unequal society. Social and cultural conditions may differ among racial groups, on average, and these differences may lead to environmental effects such as chronic stress and unequal access to goods and services, including healthcare and nutrition. These inequities can affect environmental risk for complex diseases and/or potentially interact with genetics to affect risk". All these notes are relevant from a societal point of view. The remark about stress generating a health risk certainly deserves to be considered (see above: "Non-genetic phenotypic differences between geographical populations?"). However, they mislead the reader in this debate, which focuses on *the biological concept* of race. Obviously, with such a presentation of the problem, the disqualification of the concept of race, from both a biological and moral point of view, can only be total. It can be noted that in the same article, the definition of "ancestry" perfectly aligns with the concept of subspecies or geographical race (see further: "The concept of subspecies or geographical race = ancestry groups").

This fallacious opposition between "race/ethnicity," which would be sociopolitical, and "genetic ancestry" or "ancestry group," which would have biological validity, is also put forward by Khan et al. (2022), by Jordan (2021), by National Academies of Sciences, Engineering, and Medicine (2024), by Wright et al. (2022), and by Zou et al. (2023). It is echoed by an editorial in *Nature Human Behavior* (editorial, 2022a), even though in this context, "genetic ancestry" actually covers the concept of subspecies or geographical race. Thomas and Peters (2025) note that, unlike race and ethnicity, genetic ancestry is directly related to genetic variation. However, these same authors temper this apparent opposition by emphasizing that race and ethnicity on the one hand, and genetic ancestry on the other, are linked, and sometimes *strongly correlated*, although they still recommend distinguishing the two concepts, which do have different definitions. Nevertheless, in their excellent review on the differential distribution of cancers among human geographical populations, Thomas and Peters (2025) largely take race and ethnicity as units of analysis ("proxy").

Far from this rather consensual approach, Wojcik (2025) asserts that "there is broad consensus among researchers that social constructs of descent-based identity, such as race and ethnicity, do not correspond to genetic groups." This "consensus" is questionable, to say the least (see Figures 4–8 and the many studies cited at: (https://drive.google.com/file/d/1qWNbPeyBCwDL9CgzoOMrHguFm9MeX4f2/view)

In the same vein, here is the definition (https://www.genome.gov/genetics-glossary/Race) put forward by the National Genome Research Institute, of the National Institutes of Health (NIH), an entity that is supposed to be a temple of biology and genomics: "Race is a social construct used to group people. Race was constructed as a hierarchal human-grouping system, generating racial classifications to identify, distinguish and marginalize some groups across nations, regions and the world. Race divides human populations into groups often based on physical appearance, social factors and cultural backgrounds." The only (weak) concession to biology is "physical appearance." The Wikipedia article on human races (Anonymous, 2022a) maintains the same ambiguity between sociology and biology.

This deliberate confusion between "sociological" race and human races as defined by physical anthropology (Coon, 1962; Vallois, 1976) and population genetics (Ayala and Kiger, 1984; Dobzhansky, 1962a; Hartl, 1985; King and Stanfield, 1990; King et al., 2012; Mayr, 1963, 2002) is reminiscent of the confusion between *sex*, a biological concept, and *gender*, a sociological notion (Coyne and Maroja, 2023).

Other authors (Brett and Goodman, 2021) do not operate this total disconnection from biology, but also disqualify the concept of race by accusing it of mixing skin color, ancestry (biology) with culture and religion (sociology), whereas the concept of human race, as outlined above, has been clearly defined on a purely biological level by physical anthropology and population genetics. The same confusion is present in Graves et al. (2025), who put forward a definition of race as "arbitrarily using physical, geographical, cultural, and linguistic criteria," thus sidestepping the very clear definition of population genetics and physical anthropology.

In discharge of this strategy, it is useful to note that this "sociological" meaning of race, while turning its back on the very clear definition of classical physical anthropology and population genetics, does, however, quite well matches the vision of race held by many non-specialist people.

The Continuum Versus Trees and Discontinuities Debate

In the context of this debate, many authors assert that human geographical populations constitute a genetic continuum, within which it is arbitrary to establish clear boundaries. However, this impression of a continuum is largely reinforced by taking into account populations resulting from recent interbreeding (historical era).

Another argument put forward by scientific antiracism is that human geographical populations constitute a perfect continuum, without any precise boundaries. This is called a clinal* distribution. Under these conditions, it is therefore absurd to seek to draw boundaries in this continuity. This continuum argument was put forward by Lewontin, in his foundational article of 1972, and by Edgar and Hunley (2009). More recently, Last (2012) in his critique of the book: "The 10,000-Year Explosion" (Cochran and Harpending, 2009), brought it back: "Genetic diversity is best represented as an unbounded continuum."

The picture of a continuum depicted by Lewis et al. (2022) seems confirmed by the Figure in the corresponding article. However, if one is curious enough to go back to the original article from which this data is derived (Belbin et al., 2021), a careful examination of Figure 1C of this latter article shows that the impression of a continuum is entirely due to the "Hispanic/Latino" sampling. It is well known that this community in the United States presents a whole range of interbreeding between Europeans, Africans, and Native Americans, the proportions of these three components varying greatly from one "Latino" community to another: more African ancestry on the Colombian coast, more Native American component in Bolivia. It is therefore not surprising that "Hispanic/Latinos," showing a wide range of degrees of interbreeding between Africans, Europeans, and Native Americans, form the "bridge" with everything else. Otherwise, in Figure 1C of Belbin et al. (2021), the major geographical groups, namely "African American/African," "East Asian," "South Asian," "European American," "Native American," appear very well differentiated within this sample from New York City. This result is all the more noteworthy because the identification of the different populations was based on the "self-reported ancestry* (SRA*)" system, i.e., a method by which the genetic component of the populations is unknown a priori, the origin of individuals being based solely on self-identification.

The phenomenon of Latinos/Hispanics "filling the gaps" was reiterated by Gouveia et al. (2025). These authors indeed note that the gaps between the major geographical groups (Asians, Africans, Europeans, Native Americans) are largely filled by the Latino group. Once again, this is an expected result, since this group is known to be the result of recent interbreeding (historical era) between Africans, Europeans, and Native Americans. In Figure 1C of Gouveia et al. (2025), if the Latino group is subtracted, the overlap between Africans, Europeans, and Asians is not notable. It is the Latinos who "form the bridge."

The American anthropologist Frank B. Livingstone was one of the most ardent defenders of scientific antiracism and, as such, crossed swords with Theodosius Dobzhansky (1962b) who, despite a recent assertion (Yudell et al., 2016), never adhered to the "major paradigm shift" even though his concept of biological race evolved over time (Gannett, 2013). In Livingstone's words (1962), "There are no races, there are only clines." This author therefore reiterated the continuum argument.

Tishkoff and Kidd (2004) reluctantly admit that the evolutionary trees of the human species developed from genetic markers correspond to the major geographical groups, but then add, somewhat contradictorily: "Although populations do cluster with broad geographic regions, which generally correspond to socially recognized races, the distribution of genetic variation is quasicontinuous in clinal patterns related to geography." In other words, genetic variation reliably distinguishes Africans, Europeans, Asians, Oceanians, but it is "continuous." Figure 4 of their article contradicts this assertion of a continuum and clearly separates the different human geographical populations. It is interesting to note that African Americans, who have an average of about 20% European genetic contribution, clearly group into the African cluster (grouping).

The continuum argument has also been taken up by Brown et al. (2025), by Quintana-Murci (2021) and by the late French geneticist Axel Kahn, interviewed by Villard, Grauby-Vermeil and Vidard (2015).

One might object to this continuum argument that drawing more or less arbitrary boundaries in many natural phenomena is very commonplace and is an indispensable requirement to try to establish some order within a teeming real world. I give a few examples below:

- In the south of France, at what precise boundary does one pass from the "Mediterranean" climate to the "Mediterranean-influenced" climate that prevails in my Cévennes mountains foothills?
- At what precise fraction of a second does day turn into night, a child become an adolescent, an adolescent an adult, an adult an old person? Yet, these entities are very real and no one would think of regarding them as Figments of the imagination or considering them interchangeable.
- The spectrum of colors is a *perfect continuum*. Colors only differ from each other by the wavelength of light. There is no strict boundary between blue and green, or between green and yellow. Arbitrary segments based on wavelength are established to classify the different colors. And yet, colors are not a Figment of the imagination. On the French flag, there is blue and red, and no green.

According to Alain de Benoist (1979), "to claim that races 'do not exist' on the pretext that there are a quantity of intermediate types between them amounts, not only to denying the obvious, but also to only granting the status of existence to absolute metaphysical entities."

According to the continuum argument, the approach of using genetics to define races is like trying to slice soup: "You can cut wherever you want, but the soup stays mixed" (Rotimi, 2004). The examples cited below show what should be thought of this argument.

In fact, human genetic variability, if one takes a randomized sample with representative numbers for each geographical population, does not show the image of a continuum: there are indeed "gaps", discontinuities, which correspond to the major geographical obstacles: Sahara, Himalayas, oceans. This is the classic "SHO" argument (Maglo et al., 2016). This had already been observed by classical population genetics (Dobzhansky, 1962a) and has been amply confirmed by modern genetic markers such as microsatellites (Rosenberg et al., 2002, 2005). The least that can be said is that Rotimi's (2004) "soup" has some nonnegligible lumps (Tibayrenc, 2017b). According to Rosenberg et al. (2002, 2005), human genetic variability shows two aspects: the main continental groups show a cluster distribution, within which the distribution tends to become clinal.

Kind of Solomon's Judgment was formulated by Shiao et al. (2012), who qualified this picture as "clinal classes," meaning classes that present a certain discontinuity between them, but within which the variation would be clinal. This altogether very diplomatic opinion, however, drew harsh criticism (HoSang, 2014).

Biological Evolution has Stopped: The Human species Undergoes only Cultural Evolution

This conviction is shared by many, particularly by leading scientists (Gould, 2000; Lawler, 1978; Mayr, 1963).

> *The concept that human biological evolution stopped millennia ago would imply that there are few innate differences between geographical populations. However, it is contradicted by much data.*

Obviously, such a view reduces the probability of innate differences between geographical populations, which aligns with the vision of "scientific antiracism." However, much data argues in favor of the permanence of recent, and even very recent (historical era), or even *current*, biological evolution in the human species.

For example, our species continues to be fully subjected to the selective pressure caused by transmissible diseases. Let's take the case of malaria*, particularly its most malignant form caused by the parasite *Plasmodium falciparum*. Every year, particularly in sub-Saharan Africa, this disease continues to cause the death of thousands of young children (Esoh, Yalcouyé and Wonkam, 2025). These children will never reach reproductive age, and therefore, will never transmit their genes to their offspring. In contrast, the survivors will be able to do so. This is the very statement of natural selection: individuals who cannot transmit their genes to their offspring are counter-selected. Therefore, this selective pressure continues to operate today, and the same is true for many transmissible diseases. This explains why, statistically, African children are more resistant to malaria than European children (Gentillini et al., 2012).

Since the distribution of transmissible diseases is very different from one region of the globe to another, this constantly operating selective pressure tends to differentiate geographical populations from each other, and this continues even today. This problem of recent, or even historical, human biological evolution will be developed in Chapter 3 of this Part I ("a few illustrative particular cases").

Do Human Races Exist? Then list Them!

> *The delimitation and number of races vary considerably from one author to another, which would demonstrate the futility of this concept.*

This is yet another argument repeatedly used by scientific antiracists, who point out that the number of races identified by different authors varies considerably, which, in their eyes, shows that the concept of race is inept (Foucart, 2013). Raff

(2014), in his critique of Nicolas Wade's book "A Troublesome Inheritance" (2014), reiterated the same argument.

This refers to a very old debate in taxonomy, already evoked by Darwin, that of the "lumpers" and the "splitters". The former tend to describe "good, big species," while the latter churn out bursts of small species separated by tenuous differences, which can be good for the publication list. With these "splitters," of course, the risk of exaggerating the description of new species is great. The risk obviously applies to human races. Thus, one author goes so far as to describe 200 races! (Dobzhansky, 1962a).

A common misconception is to think that races are a recent creation, elaborated to support slavery and colonialism. This may be true for racism, but the act of distinguishing races dates back to the highest antiquity, as noted by Sarich and Miele (2004). The ancient Egyptians distinguished 4 major racial divisions: Egyptians, Asians, Libyans, and Nubians (Anonymous, 2021b).

Johann Friedrich Blumenbach (1752–1840, German physician and anthropologist, is the author of a classification into 5 main races, which has been more or less corroborated by many authors. These races are: Caucasian (= Europeans), Mongolian (= Asians), Ethiopian (= Africans), American (= Native Americans), and Malayan. It should be noted that Blumenbach defended the concept of the equality of races, all equal in aptitudes according to him.

Carl Linnaeus (1707–1778), the Swedish naturalist at the origin of the taxonomic system that bears his name, in his 1740 work: *Systema Natura* (2nd edition) distinguished 4 races, which he called "varieties": "Europaeus albus" (white Europeans), "Americanus rubescens" (red Americans = Native Americans), "Asiaticus fuscus" (yellow Asians), and "Africanus niger" (black Africans).

Closer to us, the American anthropologist Carleton Coon, author of the famous work "The Origin of Races" (1962), began by distinguishing 5 main races, which partially overlapped with those described by Blumenbach: Caucasoid Race (Europeans), Congoid (Africans), Capoid (Hottentots of South Africa), Mongoloid (Asians), and Australoid (Australian Aborigines). Coon is at the origin of a theory now invalidated, at least partially, on the origin of these 5 races (see in this same chapter: "The origin of human races"). Subsequently, this author and his students sought to refine their classification system, describing about 30 different races (Dobzhansky, 1962a).

French anthropologist Henri Vallois, in the famous "Treatise on Zoology, Anatomy, Systematics*, Biology" by French zoologist Pierre-Paul Grassé, described human races as subspecies of the species *Homo sapiens*, and as such gave them "trinames," the official nomenclature for subspecies in Linnaean taxonomy: *Homo sapiens australosicus* (Australian aborigines), *H. sapiens albus* (European whites), *H. sapiens asiaticus*, and *H. sapiens afer* (Africans). Actually, according to Vallois, these subspecies corresponded to "groups," which could be subdivided into 27 races.

The Centers for Disease Control (CDC), a federal institution located in Atlanta, Georgia, where I worked for a year (1995–1996), is the main American organization in charge of public health, with a research mission as well. The CDC promote a racial classification (Burchard et al., 2015) similar to that used by American census institutions and the Office of Management and Budget: "African American",

"Black", "Latino", "Hispanic", "Asian", "Asian American", "Native American" "American Indian", "Alaska Native", "Native Hawaiian", "Pacific Islander" and "minority". Such a classification obviously contains an arbitrary component and has been repeatedly criticized by proponents of scientific antiracism for this reason. It would, however, be wrong to consider it totally disconnected from biological reality. It can be recalled that the categories "Latino" and "Hispanic" are very heterogeneous, depending on the country of origin of these populations. The populations of Latin America have three main components: European (mostly Iberian), African (from the slave trade), and Native American or Amerindian. The contributions of these three populations vary from country to country. In Bolivia, as we saw in the introduction, the Amerindian component is dominant, and populations of African origin are very much in the minority. Conversely, on the Colombian coast, people of African descent are numerous. The fact remains that the overall biological composition of Latinos/ Hispanics is, *on average*, very different from that of Americans of European or African descent, and from Amerindians.

Leroi (2005), in the columns of the *New York Times*, notes that the characterization of geographical populations by many genes reliably produces a classification reminiscent of the major races of classical physical anthropology: Europe, East Asia, Africa, America, and Australasia.

What can be concluded from this overview of racial classifications? It is undeniable that there are profound divergences regarding the number of races distinguished by different authors. But this relates to the level of precision these authors wish to achieve: to be satisfied with the main continental groups which, themselves, seem very robust, or to describe human diversity more finely. Similarly, a painter will not be satisfied with a crude classification of colors into blue, green, red, etc., and will specify: ultramarine blue, Veronese green, carmine red, etc. This described number of races actually depends on the purpose of the study.

> *The described number of races depends above all on the purpose of the study (a problem of nomenclature). The existence of races would, however, be an objective biological fact. The major continental groups seem to be very robust, and are found, more or less, in the various classifications.*

As Dobzhansky (1962a) notes, the *existence* of races is an objective biological fact, but the number of races one describes is a matter of convenience, a problem of nomenclature. The fact that this number fluctuates from one classification to another is absolutely not an argument, according to this author, for decreeing that the concept of race is invalid.

Human Races are Difficult or Impossible to Define from a Biological Point of View

This argument of scientific antiracism overlaps with those of the continuum and the number of races. According to it, racial categories are not clear-cut, are very heterogeneous, have fuzzy boundaries, vary from author to author, and simplistically describe the genetic diversity of our species.

> *The argument that human races are a difficult concept to define applies even more to a number of diseases, particularly autoimmune and psychiatric conditions.*

Such a hazy concept (human races) should therefore be rejected. This argument could just as easily be applied to a number of categories that we use, particularly in the medical field. So-called "autoimmune" diseases (e.g., systemic lupus erythematosus, or dermatomyositis) are an extremely fuzzy category, with still very poorly understood causes, very misleading symptoms, and transitional forms from one pathological framework to another. And yet, the physician is constrained to the least imperfect diagnosis possible, because this will determine prognosis and treatment.

When considering mental disorders, it is even worse. Their description changes drastically from one era to another, and consequently, the prognosis and treatment attached to them. It is relevant to underline that when I was a medical student in the 70s, homosexuality, whether masculine or feminine, was considered a severe disorder, a serious perversion, and this, by the most official psychiatry, for example in the reference manual of Ey, Bernard and Brisset (1960), or in the textbook for medical students by Van Laere (1961), and even in popularization books (Daco, 1973). It is needless to insist that the current picture is very different. However, to my knowledge, no major discovery has been made in this area, either in biology, genetics, or brain imaging. It is not science that has changed, it is society's view of homosexuality.

If one is curious enough to consult the famous "Diagnostic and Statistical Manual of Mental Disorders" (DSM) issued by the American Psychiatric Association (Hyman, 2014), one realizes, with successive versions (we are at the 5th), that the status of a number of mental disorders has radically changed. This could give a strong impression of arbitrariness. And yet, this effort of nomenclature and description is indispensable. The definition of mental disorders may be fluctuating, but the suffering of the patients is real, and calls for diagnosis, prognosis, and treatment, in the current state of medical science.

The fact remains that in comparison with these very vague entities described in the fields of autoimmune or psychiatric diseases, the 5–6 main geographical human groups ("major races") identified as early as the time of Linnaeus (18th century), appear to be models of resilience.

Human Races are Social Constructs with no Biological Basis

This line of attack, already presented (see: "Aiming for the wrong target (Straw man strategy? Sociological definition of human races partially or totally disconnected from biology)"), is the favorite mantra of scientific antiracism. The underlying concept, implicit or explicit, is that races are entities that are defined solely on historical and sociological criteria, with little or no biological reality. In its ultimate form, the concept explicitly aims to discourage any study taking into account biological facts that could distinguish geographical populations from one another (see further: "The temptation of censorship"). This position is very dominant among human science specialists and even among some biologists. Even David Reich, the famous specialist in ancient DNA, who received a significant backlash (Mary, 2018a; Nelson, 2019) after his iconoclastic and somewhat suicidal positions on the "substantial" differences (his own terms) that might exist between geographical populations (Reich, 2018 a and b), had to make a prudent semantic shift, and admit, maybe reluctantly, that human races were above all social constructs (Reich, 2018c): "Race" is fundamentally a social category—not a biological one—as anthropologists have shown." This is called a split, given that all the rest of his discourse says exactly the opposite.

Race as a largely or solely social construct is the official position of the American Association of Anthropologists (1998). It was recently reiterated by the American Society of Human Genetics (ASHG perspective, 2018). The author of this book sent a critique of this article (Tibayrenc, 2019) to the *American Journal of Human Genetics*, which had published it. The critique was courteously published by the journal and drew a very measured response. Americans have a sense of debate.

> *The argument that race is a purely sociological concept with no biological basis is accepted by many authors, including biologists.*

Many authors have defended this thesis of a purely sociological concept of race. I refrain here from citing all the articles I have found corroborating it. I report below a representative sample of this current of ideas.

As we have already seen (see: "The continuum versus trees and discontinuities debate"), an undoubted pioneer of this movement was the American anthropologist Frank B. Livingstone (1928–2005). His 1962 article "On the non-existence of human races" resonated like a cymbal crash in this scientific community. Livingstone›s argument was original and scientifically clever: instead of criticizing the concept of human races in isolation, attacking the evil at the root, he attempted to disqualify the concept of geographical races or subspecies in general, and this, throughout the animal kingdom (see further: "The concept of subspecies or geographical race = ancestry groups"). The evolutionist Theodosius Dobzhansky strongly opposed (1962b) Livingstone's views. According to him, the scientific antiracism defended by Livingstone "plays into the hands of race bigots." This pitfall raised by Dobzhansky should not be underestimated.

Another significant Figure of this "racial denialism," already mentioned, is the American geneticist Richard Lewontin, who could be described as the "pope of scientific antiracism."

Professor Albert Jacquard (1925–2013) could apply for the same title. His Fight for scientific antiracism still resonates. Among his many popular science books, one can cite the emblematic "Eloge de la différence : la génétique et les hommes" (In Praise of Difference: Genetics and Men) (1978). No offense to his (numerous) detractors, as I have already had the pleasure of emphasizing, according to my friend and co-author Francisco J. Ayala, Professor Jacquard was one of the greatest names in contemporary population genetics. Professor Jacquard's arguments against the concept of biological race will be discussed in: "The concept of subspecies or geographical race = ancestry groups."

The French molecular biologist Bertrand Jordan is another defender of this concept (2008, 2010).

For Cerdeña, Grubbs and Non. (2022), races are only social categories deliberately generated by the racist ideology of white people to assert their domination over black and "brown" (?) people. According to these authors, the use of racial categories in medicine is a "Tinderbox."

According to the historian Catherine Coquery-Vidrovitch, a specialist in Africa (cited by Kodjo-Grandvaux, 2021), "All reasonable and honest people know that races do not exist." It is not certain that this kind of peremptory statement, coming moreover from someone who does not have any expertise in biology, serves the cause of antiracism.

The French journalist Stéphane Foucart (2013) also considers human races to be "social constructs."

According to Lewis et al. (2022), racial classifications are the result of social, economic, and political processes.

According to Popejoy (2021), racial groups are defined by societies, not by genetics.

The French geneticist Lluis Quintana-Murci, in his recent work: "Le peuple des humains" (The People of Humans) (2021), claims that: "The concept of race has no place from a biological point of view in humans, and 'race' is essentially a cultural construct."

According to Schwartz (2001), "race is a social construct, not a scientific classification. There is only one race, the human race." The author, carried away by his momentum of scientific antiracism, commits a major taxonomic blunder here, because *Homo sapiens* is not a race, but a species. This semantic confusion is moreover taken up by many participants in this debate, particularly in the reader forums of major media outlets.

The geneticist and Paris councillor Dominique Stoppa-Lyonnet (2015) asserts that race is a cultural, not a biological, creation.

Wade (2021 a and b) asserts that society, and not biology, defines races and denounces the erroneous notion that race is biological.

Taguieff (1988) recalls that in France, the three champions of this concept are the geneticist Albert Jacquard, the Nobel Prize winner François Jacob, and the biologist Jacques Ruffié. According to these scientists, if human races have no biological reality, racism is inept. Taguieff, in his most recent essay: "race? Un mot de trop" (race? A word too many) (2018) cites the famous 1942 work by the British anthropologist Ashley Montagu, with the most telling title: "man's most dangerous myth: the fallacy of race" (Another possible translation: the sophism of race). According to the same Montagu, the simple act of talking about race is racism (Wade, 2014).

As already explained (see above: "Non-genetic phenotypic differences between geographical populations?"), some authors (Brown et al., 2025; Pontarotti, 2021) defend a more subtle variant of this concept of race as a cultural creation: they claim that the real biological differences that can be observed, for example, between African

Americans and white Americans are not due to intrinsic genetic differences, but are the result of the discrimination suffered by the first group, with this discrimination (moral humiliations, systemic racism, polluted environment, poor diet, lack of access to care) affecting the organism through epigenetic mechanisms.

> *The hypothesis according to which racism leads to unfavorable phenotypic modifications in African Americans via epigenetics is based on a typological view, suggesting that all members of this community undergo uniform discrimination.*

One can note that such a view implies a typological conception of racial groups: it presupposes that all African Americans undergo uniform discrimination, and that all White Americans benefit from a homogeneous absence of discrimination. This is absurd. Even if the African-American community remains statistically disadvantaged, there are African-American classes in the United States that are quite affluent, reaching the highest strata. And many poor white people, particularly in the Southern states, endure extremely harsh lives. That discrimination has epigenetic effects is quite probable. But this should not exempt us from also seeking classic genetic mechanisms likely to explain certain medical differences between racial groups.

The Concept of Human Race must be Rejected Because it is Dangerous

This is an argument that cannot be dismissed with a flick of the wrist. The "criminal record" of the race concept—a concept that has been widely used by racism, colonialism, slavery, and Nazism—is of course horrifying. Many authors have therefore considered that this notion of human race must be firmly rejected given the risk of resurgence of past crimes. It is certain that if one has the unhealthy curiosity to browse certain current websites of the "identitarian" or "supremacist" movement, one can easily see that the Nazis have spawned successors.

Jordan (2008) thus considers that race is not only a poorly defined notion but is also responsible for many atrocities. It is therefore much better to abandon it.

Let us cite Cavalli-Sforza (2007): "I am inclined to dismiss the word 'race' because of its connection with the odious episodes of racism with which we are continuously confronted." Let us note here that the author speaks only of eliminating the term "race," but not the underlying biological concept.

According to the well-known and late Albert Jacquard, the concept of race should be rejected because it is useless and, above all, dangerous (cited by Taguieff, 1988).

According to Taguieff (2018), this is a very widespread position among "scientific antiracists."

According to Wolpoff (2009), the rejection of the concept of human race is due to several reasons, one of which is political.

The famous "ancient DNA hunter" Svante Pääbo (2016) concurs: "On the other hand, I do not claim to Fight racism because for me it is not a scientific question, but rather an ethical and political stance."

The highly political aspect of the question was confirmed by the fact that on June 27, 2018, French deputies unanimously voted to remove the word "race" from the constitution (Equy, 2018). However, voices were raised against potential perverse effects of this decision, particularly the possible weakening of the antiracist struggle if the concept of race is evacuated.

This conduct of avoiding a "word too many" (Taguieff, 2018) was qualified by Alain de Benoist (1979), the leader of the French School of thought "New Right," as "semantophobia."

The rejection of the term "race," which is deemed dangerous, is very understandable, but the problem then arises of substituting a perfectly satisfactory term for it.

Notwithstanding this ironic note, the discomfort with such a historically and politically loaded term cannot be said to be contemptible or trivial. But on the one hand, there is the difficult semantic question of finding adequate terms that can substitute for the word "race" (see Part II, Chapter 2). On the other hand, we touch here upon one of the central questions of this book: to what extent can or should moral and ethical considerations interfere with a purely factual and objective cognitive process—a scientific process? Is it acceptable, or even desirable, to censor an entire field of scientific research because some have made criminal use of it, for fear that this might reoccur?

Treating Humans as a Specific Case

Ceci and Williams (2009) carry a message that appears more measured than those mentioned in the previous paragraph. Although they do not believe in genetically-based differences between geographic groups (= races), they consider that reducing scientists to silence resembles religion more than science. However, they add this curious remark that still smacks of censorship: "Perhaps such research should be forced to pass a higher cost-benefit threshold before publication." In other words, research on human geographic diversity would be subject to specific, more stringent acceptability criteria than comparable research on Californian newts or Galapagos finches.

The approach of treating humans as a special case from a biological point of view is highly criticizable: one cannot elaborate one biology for humans and another one for the rest of the living kingdom.

In my view, this is unacceptable. There must not be a biology for humans on the one hand, and a biology for newts and finches on the other hand. Biology's methods and requirements must be uniform. Making a special case of humans is the sin of anthropocentrism, a possible resurgence of the religiously rooted thought that man is a separate being, of divine origin and of more or less immaterial nature.

This is a perfect approach for having a biased and false view of the biology of our species. It is also a convenient subterfuge to avoid certain comparisons with our animal friends, comparisons that could prove embarrassing for the proponents of one ideology or another. A classic sophistical trick: a particular case is isolated to conceal the fact that it is, in reality, very generalizable.

Statistical Artifices of Scientific Antiracism {Too Many Mixed-Race Individuals, Too Much Africa, Filling The Gaps, Fabricating The Continuum}

The approach of those who desperately want to see races can be summarized: they will endeavor to emphasize the differences between geographic populations. The racist approach, which aims more specifically to hierarchize human groups as inherently "unequal," will insist more particularly on supposed superiorities and inferiorities in the cognitive domain. Racists have muted the supposed physical (athletic) superiorities of the elected race (theirs in general) since the historic thrashing inflicted on the handsome, blond, blue-eyed, and sufficiently dolichocephalic "Aryan" athletes by the African American champion Jesse Owens (Figure 9) (four gold medals) at the 1936 Olympic Games in the presence, and to the fury, of the Führer Adolf Hitler.

Failing to establish an indisputable athletic superiority of the "elected race," proponents of the hierarchization of geographic populations then focus on cognitive traits. This point will be developed in Part II: Racial Theory. We will simply say here that this problem of innate cognitive differences constitutes the very essence of racism, or of antiracism.

> *The racist ideology has long abandoned the ground of a potential athletic superiority of the "superior" race and focuses on supposed innate cognitive inequalities.*

According to an approach of perfect symmetry to that of the proponents of racial hierarchy, the detractors of the concept of biological race will do everything possible to minimize the differences and boundaries that might exist between geographic populations.

> *The strategy of the proponents of "scientific antiracism" is to minimize the differences between geographic populations through various sampling procedures.*

We have already set out the "continuum" argument (see: "The continuum versus trees and discontinuities debate"), which is the main angle of attack for minimizing the differences between geographic populations within the framework of the typological concept of race. Everything is done to reinforce it: the sample is enriched with mixed-race individuals (Lewis et al., 2022), which are intermediate by nature, and numerical emphasis is placed on "suture" (intermediate) populations. Another strategy consists of giving excessive numerical importance to populations of African

Figure 9: The African American champion Jesse Owens, four-time gold medalist at the 1936 Berlin Olympic Games. Chancellor Adolf Hitler, angered by the fact that the Berlin crowd warmly cheered the Black champion, refused to congratulate him.

origin, which has the effect of "crushing" the other populations and minimizing the differences between the various groups (Carlson et al., 2022). This thus turns away from the golden rule of statistical sampling: it must be representative of the overall population under study.

The Concept of Subspecies or Geographic Race = Ancestry Groups

The problem of subspecies in general has already been raised above (see: "The subspecies or geographic race: a concept of importance for this debate").

The proponents of "scientific antiracism," as we have seen, have concentrated their attacks on three concepts of human races: typological, phylogenetic, and sociological. This angle of attack can be seen as a voluntary, unvoluntary or even unconscious, "straw man strategy": to disqualify a concept, one attacks it for what it is not. Indeed, what is clearly apparent is that these three concepts have never been used, to my knowledge, either in population genetics (Ayala and Kiger, 1984; Dobzhansky, 1962a; Hartl, 1985; King and Stanfield, 1990; King et al., 2012; Mayr, 1963, 2002), or in physical anthropology (Coon, 1962; Vallois, 1976). These disciplines have always unambiguously equated human races with "subspecies" or "geographic races."

It is interesting to note that the French historian of science André Pichot, in the progressive daily newspaper *Le Monde* in 1996, gave human races a definition very close to this concept of subspecies.

> *Population genetics and classical physical anthropology equate human races with subspecies or geographic races. "Scientific antiracism" never refers to this model.*

Subspecies are a taxonomic category whose validity is disputed by some authors (Lherminier and Solignac, 2005). They correspond to geographic variants of a given species. Despite controversies about the validity of the concept itself, it is officially recognized by the International Code of Zoological Nomenclature (ICZN) (https://code.iczn.org/?frame=1). A subspecies identified by a taxonomist is labeled by a *triname*. For example, the Zebu and its cousin the European cattle are named *Bos taurus indicus* and *Bos taurus taurus*, respectively (see Figure 1). If Neanderthal* man was not considered a distinct species from modern man, he would be made *Homo sapiens neanderthalensis*.

> *The taxonomic category subspecies, or geographic race, is not accepted by all biologists. However, it is officially recognized by the International Code of Zoological Nomenclature.*

Many authors have provided converging definitions of the concept of subspecies or geographic race.

Regarding this concept in the living kingdom in general, not just the human species, Ayala and Kiger (1984) give the following definition: "Races are populations that possess somewhat differentiated gene pools. The differences between races must involve the gene pool as a whole, and thus allele frequencies at many loci. Differences at a single locus or a single trait may serve as indicators of overall genetic differentiation, but they do not alone constitute a sufficient basis for distinguishing races." This definition highlights an important point: subspecies or geographic races are not characterized on the basis of isolated traits, but *on a set of traits*.

> *Subspecies or geographic races must be validated on the basis of the whole of their characters, not on isolated traits.*

This point is often used as an angle of attack by "scientific antiracists," who note the apparent discordance between one trait and another in the case of human races: this is the "discordance of traits" argument. We will return to this. The definition by Ayala and Kiger (1984) invalidates the discordance of traits argument.

The anthropologist Carleton S. Coon, an emblematic "race realist," gives the following definition: "Subspecies: a major division of a species, constituting a geographic race. Race: a general term referring to distinct genetic divisions of a (given) species." (Coon, 1962).

According to Mayr et al. (1953), a subspecies or geographic race is "a geographic aggregate of local populations that differs taxonomically from other similar divisions of the species. A subspecies is a geographic race sufficiently different taxonomically to warrant a distinct name."

Subspecies, or geographic races, are geographic variants of the same species. They remain potentially inter-fertile.

To summarize: a subspecies, or its synonym the geographic race, is a subdivision of a given species that differs *statistically* from the other subspecies of the same species by the frequency of certain phenotypic or genotypic characters. By definition, there is no genetic isolation between subspecies of the same species. They are potentially inter-fertile if they have the opportunity to interact.

It is recalled here (see: "The continuum versus trees and discontinuities debate") that a major "scientific antiracist," the American anthropologist Frank B. Livingstone (1962), aptly attempted to "aim for the head," and, rather than specifically attacking the concept of human race, undertook to flat-out denigrate the notion of subspecies or geographic race in general, thus hoping to nip the very idea of human race in the bud. This led to an epic duel with the evolutionist Theodozius Dobzhansky (1962b), a "race realist" but not suspect of racism.

The Specific Case of Human Races has been the Subject of the Following Definitions

First, here is the definition from the French dictionary "Petit Larousse", 2009 edition. It is doubtful that much has changed since then: "A category of biological classification and hierarchization of different human groups, *scientifically aberrant* (my emphasis), whose use is at the origin of different racisms and their practices."

In the francophone reference book on tropical medicine directed by Professor Gentillini (4th edition, 1986), the following definitions are found (page 695): "race (defined by a set of common hereditary characteristics); ethnie (characterized by a common linguistic, cultural, political background...)." We will return to this pertinent distinction between race and ethnie. It is worth noting that in a later edition of this work (6th edition, 2012), all reference to the concept of race was removed.

As we have seen, major authors in population genetics (Ayala and Kiger, 1984; Dobzhansky, 1962a; Hartl, 1985; King and Stanfield, 1990; King et al., 2012; Mayr, 1963, 2002) equate human races with subspecies. The same is true for physical anthropology (Coon, 1962; Vallois, 1976).

The American anthropologist Ernest A Hooton (1926) gives the following definition of human races: "A major division of humanity, characterized by the sharing of a certain combination of characteristics derived from their common descent and constituting a vague physical background, generally more or less obscured by individual variations and best described as a composite image."

The population geneticist Daniel Hartl (1985) gave this almost poetic vision of human races: "By way of analogy, the human gene pool is composed of a number of smaller pools—puddles—like the water puddles on a gravel street after a heavy rain, all interconnected in tiny rivulets through which material flows from one to the next. If one examines the puddles closely, one sees that they are not identical. But the differences are quantitative, not qualitative. The puddles are distinguished by how much or how little they have of each component and not by whether or not they possess the constituent. No puddle is completely isolated, so whatever one puddle has in abundance travels through the rivulets to the others. Because of this structure, the puddles form one large, interconnected unit, but a unit with local differences. The puddles in the analogy represent races—groups whose allele frequencies differ from those of other groups."

The famous Dictionary of Genetics by King, Mulligan and Stansfied, in its ultimate 2012 edition, gives the following definition of subspecies: "geographically and/or ecologically defined subdivisions of a species with distinctive characteristics; cf. race. Race: a phenotypically and/or geographically distinct subspecific group, composed of individuals inhabiting a defined geographic and/or ecological region, and possessing phenotypic characters and gene frequencies that distinguish it from other such groups; *Homo sapiens* can be divided into five races on the basis of geographic origin: Africans, Asians, Caucasians [= Europeans], Pacific Islanders, and Native Americans". These unequivocal definitions have not changed from the 1990 edition (4th edition).

Sarich and Miele (2004) are the authors of a book with an inflammatory reputation: "Race, the Reality of Human Differences." It is important to emphasize that these scientists, particularly the first author, have indisputable competence. Vince Sarich was a pioneer in the field of "molecular genealogies," and showed that the divergence time between humans and chimpanzees, about 5 million years ago, was much more recent than paleontologists thought. Here is what these authors say about human races: "populations or groups of populations of the same species, geographically separated from other populations or groups of populations, and distinguishable from them on the basis of innate traits." This aligns with the concept of subspecies or geographic race. Sarich and Miele insist that this definition in no way implies a sharp discontinuity between races, as falsely presented by the "typological" concept of race.

Abandoning the case of professional scientists, we cite the definition of human races given by a theorist of the French school of thought "New Right," Alain de Benoist (1979). According to this author, races are "groups of individuals exhibiting different frequencies of gene distribution and expression." A definition which is, moreover, as classical as possible, and is perfectly consistent with the definition of human races as subspecies or geographic races.

André Pichot, a French historian of science (cited by Vincent, 2013), provides observations showing that he is very close to the definition of race as a subspecies, in opposition to the sophistical typological view: "By definition, race brings together individuals who have certain common hereditary peculiarities marked enough to be gathered into a group, but insufficient for this group to constitute a distinct species. Races have never been more precisely defined. They have always allowed for a certain arbitrariness, and, contrary to what is often claimed, they have never had essentialist or typological definitions. This is why it is absurd to speak of the existence or non-existence of races: they exist in taxonomy according to the definition taxonomists give of them or according to the use they make of them. Furthermore, whatever geneticists may think, there are races in the world we live in," he continues. "For we do not dwell in a world of genes, but in a world of men, and the latter is irreducible to the former."

Lala and Feldman (2024) consider that human races do not show sufficient criteria to be considered subspecies. These authors establish the threshold at an *F*st of 0.25. This criterion is entirely arbitrary and has never, to my knowledge, been adopted to validate the concept of subspecies.

What should be retained from this long exposition on subspecies, geographic races, and the specific case of human races? That this approach to the problem resolutely turns its back on the "typological" or "essentialist" view of human races, constantly put forward by "scientific antiracists" to defend the doxa "there are no races."

> *The assimilation of human races to subspecies or geographic races is very different from the typological view of races.*

Nowhere in this concept of subspecies or geographic race do the typological characteristics of homogeneous, strictly separate races appear. Furthermore, this definition given by population genetics and physical anthropology is much clearer and more precise than the fluctuating definitions originating from sociology.

> *One cannot say <u>in absolute terms</u> that human races are subspecies, nor that they are not. It is only a <u>model</u>, inherently debatable. However, its predictive power is satisfactory.*

One cannot absolutely assert that the major geographic populations of *Homo sapiens are* subspecies (no more than one can absolutely claim that they are not). Such a proposition is only a *model*, which is far from achieving consensus among specialists on the issue. Some of my colleagues, who are highly competent in population genetics, resolutely reject it. There is no consensus on the matter, just as there is no consensus around the proposition: "human races do not exist." One can, however, assert that the subspecies or geographic race model has a satisfactory predictive power for anticipating the distinctive biological properties of these geographic populations. This predictive power is infinitely better, in any case, than the one of the "Move along, nothing to see here" model, which will be discussed later.

The "Post-Genomic Surprise": Rehabilitation of the Biological Concept of Race? End of the "Enchanted Interlude," at the cost of "Soft Landings." Manichaeism: Erroneous dichotomy Nature vs. Nurture

"The Golden Age of Scientific Antiracism"

The pioneering data from population genetics in the 1970s, led by Richard Lewontin, appeared as a godsend for "scientific antiracists," especially those who constantly highlighted the "typological," or "essentialist" concept of race.

> *Population genetics of the 70s seemed to radically invalidate the concept of human races, especially within the typological view of this concept. This was the "Golden Age of Scientific Antiracism."*

According to this typological concept, races were indeed supposed to appear as fairly homogeneous entities, separated from each other by sharp boundaries. Let us reiterate that such a picture was never defended by classic anthropology (Coon, 1962; Vallois, 1976), nor by population genetics (Ayala and Kiger, 1984; Dobzhansky, 1962a; Hartl, 1985; King and Stanfield, 1990; King, Mulligan and Stanfield, 2012; Mayr, 1963, 2002). It is certain that the image of human populations presented by Lewontin (1972) was the exact opposite of the typological model: highly heterogeneous geographic populations, with very blurred boundaries, and above all, intra-population genetic variation greater than inter-population genetic diversity (the "*F*st argument"). It was therefore only too easy to use it to disqualify this typological model, falsely presented as the concept in use for defining human races.

This "Golden Age of Scientific Antiracism" was dubbed an "enchanted interlude" (Bonniol et al., 2021). "Irrefutable" evidence had finally been obtained that human races were a purely sociological concept, lacking any biological basis. However, despite this argument from authority, many scientific authors, especially in the biomedical field, persisted, and still persist, in comparing Africans, Europeans, and Asians in their research, and even, for quite a few of them, to unambiguously use the "word too many" (Taguieff, 2018) of race. This is attested by numerous scientific publications: (https://drive.google.com/file/d/1qWNbPeyBCwDL9CgzoOMrHguFm9MeX4f2/view) (https://drive.google.com/file/d/1pffskqiM3WuPeD7gLIWYsSiXyTdwoeso/view).

Starting in the 90s, the limits of this strategy for eradicating racism through supposedly definitive scientific arguments were being reached. The public, and even many scientists, did not follow. The "abrupt" nature of the argument from authority—"science has proven that human races do not exist"—was denounced as probably counterproductive (Klarsfeld and Vernier, 1998).

The Caveat of Genomics

The immense advances in genomics have considerably complicated, and at the same time, clarified the picture. Overall, these new data have confirmed the results of population genetics from the heroic era: the major geographic populations (Europe, Africa, the Americas, Asia, Oceania) are highly heterogeneous genetically, and show no sharper boundaries than in Lewontin's time.

> *The discovery of low-frequency and rare variants weakens the* F*st argument.*

But the discovery of low-frequency and rare genetic variants has put a damper on the *F*st "knockout argument." As we have seen, these variants tend to be population-specific, which, by definition, strongly increases the *F*st value between these populations. This is all the more important because these variants are often involved in phenotypic expression, in other words, in phenotypic differences between populations. Furthermore, the use of high-resolution genomic variants (SNPs, routinely used in tens or hundreds of thousands) allows for a greater congruence than ever with the "classic" geographic divisions (Figures 6–8), as was already the case with the "genetic jalopies" of the heroic era (Figures 4 and 5). Above all, the study of pathological traits has fully confirmed the imperative necessity of distinguishing geographic populations in biomedical studies (see Chapter 2: Geographic Populations and Medicine).

Absence of Scientific Consensus on the Notion of Race

The consensus on the absence of a biological basis for human races, falsely reaffirmed by Brown et al. (2025), Graves et al. (2025), Khan et al. (2022), and Lala and Feldman (2024), has never been absolute. In 1973, Theodosius Dobzhansky, one of the fathers of the "synthetic theory of evolution*", stated that "it is the contention of a small but vociferous group of students that humankind is not differentiated into races" (p.52). Vociferous, certainly, but less and less small.

> *Contrary to what is often claimed, there has never been a consensus on the existence or non-existence of human races. On the other hand, there is a clear trend toward avoiding the use of this term in isolation in scientific publications.*

In 2004, in the special issue of *Nature Genetics* published on this subject, the viewpoint of the different contributors on this question was very far from homogeneous. For example, Cho and Sankar (2004) believed that the use of the race concept risked discrediting all genetic research. In the same issue, Collins (2004) had a more moderate view. He insisted on the fuzzy nature of the race concept, while acknowledging that there could be some association between self-identified race and gene frequencies, and that the concept could be useful in medicine. According to Jorde and Wooding (2004), "races are not biologically meaningless." According to these authors, if a sufficient number of genetic markers were used, the congruence with classic anthropological classifications was quite good. They noted that the *F*st statistic treats each gene individually, and thus fails to recognize this congruence, which we have already noted (see "Lewontin's fallacy"). They maintained that the race concept could be useful in medicine but stressed the continuum distribution of genetic variability. The good correspondence between genetic markers and self-attributed race was confirmed by Mountain and Risch (2004). These authors pointed out that, faced with the results of the Human Genome Project, some scientists concluded in favor of the existence of races, while others drew the opposite conclusion. Still in the same issue of *Nature Genetics Supplement*, Parra, Kittles and Shriver (2004) observed a strong correlation between skin color, assessed not subjectively, but by spectrophotometer—a perfectly objective and quantifiable measurement—and ancestry assessed by "Ancestry Informative Markers* (AIMs*)." Such a result shows that measuring skin color can reliably estimate the degree of admixture. Patrinos (2004) stated that using the race concept led to "bad medicine and bad science." Rotimi (2004) endorsed this very negative view. This author is the source of the famous quote: "distinguishing human groups is like slicing soup." Contrary to these opponents of the race concept, Tate and Goldstein (2004) considered its use in medicine to be relevant. This opinion was shared by Tishkoff and Kidd (2004). These authors postulated an absence of difference in intelligence between geographic groups. In this article, the interesting notation was made that "all men are created equal is a moral statement, not a scientific one." This view aligns with one of the main messages of the present book, namely the essential separation between moral belief and science.

The conclusion is clear: in this special issue of *Nature Genetics Supplement*, which dates back more than 20 years, there was no consensus on the concept of race among the authors. The only consensus was the use of cautious quotation marks when the term "race" was used.

According to Caulfield et al. (2009), the concept of race will continue to be used in biomedical research for a long time. These authors insist that science cannot independently dictate the meaning of the word "race" or replace it with new terms like ethnicity or ancestry. This remark is important: what is the relevance for science of generating new terms whose meaning will be understood by an infinitesimal minority? (See Part II, Chapter 2).

In an article that drew virulent responses, Huston and Raymond (2013) strongly assert the existence of human races, while noting that they lack a precise definition. The somewhat provocative title ("Yes, Races Exist!") was not the one the authors had proposed upon submission of the article, the first author, a close colleague of mine, confided.

According to the sociologist Rogers Brubaker (cited by Taguieff, 2018), genomic differences between geographic groups are "explicitly linked to popular concepts of race." Such repeated observations have been described as the "molecularization of race" (Taguieff, 2018). However, overall, the mainstream media, particularly in France, maintain the editorial line that "races do not exist" (see for example Foucart, 2013). But it is increasingly evident that the "enchanted interlude" is tending to become disenchanted.

Soft Landings

In view of this new landscape, some authors are cautiously attempting what I call a "soft landing"—that is, in veiled terms, announcing that some nuances might be needed to the "no races" consensus.

We have already cited Klarsfeld and Vernier (1998) and their relevant warning that the abrupt argument from authority—"science provides irrefutable proof that human races do not exist"—risked being counterproductive. Put differently, the proof is not quite so irrefutable.

> *Some authors, in cautious terms, question the doxa of the non-existence of human races.*

Doron (2013) speaks of the "somewhat hasty" declarations of some scientists after the war, "which concerned only a very specific conception of race" (i.e., the typological concept of race, describing them as genetically homogeneous entities, separated from each other by strict boundaries).

Hochman (2013) states that: "Nobody in the debate actually holds the *breathtakingly mad view* (my emphasis) that 'race' is entirely divorced from biology." I am not sure that this appeased and consensual view is accurate. Many debaters, especially in the social sciences, continue to maintain that there is absolutely nothing biological in the concept of race, which is only a social construct, despite the wise recommendations of the same Hochman (2013): "Social constructionists about race can no longer simply cite Lewontin's (1972) in defence of their view. The debate has moved on."

Jordan (2021), despite being one of the current leaders of "scientific antiracism" in France, recognizes that the affirmation "human races do not exist" must be nuanced, as it is "a bit simplistic." He notes that using the term "ethnic group" in preference to "race" aims to be "more politically correct." We will return later to the fact that this widely used euphemism is not satisfactory, for both historical and semantic reasons (see Part II, Chapter 2: Semantic Puzzle: What to replace "race" with?).

Sauer (1992) proposes a formulation similar in prudence to that of Hochman (2013): "No one who argues against the race concept denies that human variation exists or claims that this variation is not systematic" {i.e., distributed in connection with the geographic distribution of populations}. Nobody? Nothing is less certain, as we have seen.

Thomas and Peters (2025), while emphasizing the differences in definition between race and ethnicity (sociological concept) and genetic ancestry (biological concept), nevertheless note that the two concepts are linked, and *sometimes strongly correlated* (my emphasis).

Winther (2018) dares to confront a taboo among taboos, that of cognitive differences between human geographic groups, by insinuating that "perhaps differences in gene frequencies between populations would play some role in cognition, physical properties, and disease." The author, who cannot be suspected of supporting hierarchical differences between populations, quickly adds that "human physical and cognitive properties are universal."

Gaydarska et al. (2024) cite a recent report by the IBC (International Bioethics Committee) that states a similar viewpoint. It merits an exhaustive quote: "Present and future scientific discoveries are likely to reveal even greater fragmentation based on geography, race, and gender, allowing the genetic classification of not only health and disease but also personal traits, such as intelligence. As a result, the risk of discrimination may increase. Therefore, almost 10 years after the IBC (International Bioethics Committee) report, it is time to seriously consider future directions and policymaking in light of the UNESCO declarations." The use of the term race without the usual careful quotation marks is noted here.

Stoczkowski (2006) practices a slightly less soft landing, since he describes the UNESCO doxa that "human races do not exist" as "outdated," and considers "that it was a mistake."

David Reich puts his foot in it by suggesting that human races might differ from each other, not only in skin color but also in cognitive properties, before backtracking.

David Reich, the famous Harvard professor, the "pope" of ancient DNA, favored us not with a soft landing, but with a real "crash landing," and even several ones, which caused quite a stir (Reich, 2018 a, b, c). It would be incongruous to suspect this distinguished academic, the prototype of the East Coast American scientist, of racism, although some did not refrain from doing so. But he seems to have very unwisely let the fox into the henhouse by asserting that differences between races were not limited to skin color, and that scientific comparisons between races should concern not only susceptibility to diseases but also behaviors and cognition. Dissatisfied with his audacity, he then stressed that human races have been separated from each other for tens of thousands of years, which has led to numerous differences in allele frequencies. These differences could therefore, according to him, concern behavior *and cognition* genes. According to Richardson (2011), this is a "potentially explosive" subject. Explosions, moreover, were not long in coming. In an op-ed in the French newspaper *Le Monde* with the inflammatory title: "Race: Genetics Facing Its Demons," Mary (2018a) reports on these chain reactions. In doing so, she repeats the perennial *F*st argument, and caricatures the biological race concept, presenting it under its typological model, with geneticists like Dobzhansky or Fisher supposedly describing "homogeneous and stable" racial entities (which they never did). Mary cites the French geneticist Evelyne Heyer for this purpose: "There are no distinct boundaries between human groups that would allow for the definition of 'sealed' categories," a classic typological picture that was never defended, let us recall, by either classic anthropology or population genetics. The article also accuses Reich of passing over in silence the possible effects of the environment on certain pathologies such as prostate cancer, which is false.

Faced with this outcry, David Reich, who is no fool, performed a prudent retreat in his ultimate article dealing with this subject (Reich, 2018c), giving all necessary assurances to the "race, a simple social construct" view. Professional assassination for ideological reasons is becoming all too common. It would have been a great pity for the scientific community to be deprived of one of its most eminent members. One can, however, regret that this indispensable debate was once again censored in the name of a regrettable scientific bigotry.

Some "soft landings" are not clearly announced but lead to perilous semantic and conceptual contortions.

Bherer (2019) cites sociologist Juliette Galonnier, according to whom race is a fiction, but racist discrimination is not. But how to combat it without identifying those who are discriminated against, given that ethnic statistics are forbidden in France? Very simple: one must rely on their appearance, in other words, on their phenotype. Those who have an "Arab face" will be categorized as Arabs, those with an "African face" as Africans. Thus, race is a fiction, but one can identify an individual's geographic origin by their facial features.

Sauer (1992), already cited, tries to temper his "soft landing": "In fact, it is systematic variation that allows anyone to estimate, with varying degrees of specificity, a person's place of ancestry from their physical features. However, to identify a person as having ancestors from, say, Northern Europe does not identify a biological race of Northern Europeans." The same remark applies as for the above-cited approach of Julienne Galonnier: there is no biological race of Northern Europeans, but Northern Europeans can be identified based on their phenotype.

Maglo, Mersha and Martin (2016) affirm that human races, which they scornfully describe as a "wastepaper basket taxon", have no evolutionary significance. However, they add, it is recommendable to take them into account pragmatically in medicine, "distinguishing reality from utility." In other words, races do not exist (they are simple "wastepaper basket taxa"), but they are medically important.

Verma et al. (2024) use a baseline sample composed of Europeans, Africans, mixed Americans, and Asians in the context of a large Genome-Wide Association Study* (GWAS*). They seem to apologize for this selection through laborious circumlocutions: "These labels are applied with the understanding that they represent broad groups of genetically similar people. While they are not intended to determine racial or ethnic identities, *they are inextricably linked to these social constructs* (my emphasis). We recognize that such categorizations, although necessary for analytical clarity, oversimplify the rich and complex mosaic of human genetic diversity."

Sophistical Manichaeism/Dichotomy Between Nature and Nurture, Between Genetics and Environment

> *Many authors artificially oppose genetic and environmental contributions to the genesis of phenotypes, particularly diseases, and falsely accuse biologists studying the impact of genetics of totally neglecting environmental factors.*

Some articles criticizing Reich's assertions accused him of neglecting the effect of the environment when he postulated a genetic substrate for prostate cancer. This perennial and fallacious dichotomy between nature and nurture characterizes, for example, the work of Graves and Goodman (2022): "Racism Not Race; Answers to Frequently Asked Questions." Just because prostate cancer has a genetic component (which seems to be well-supported) does not mean that the environment plays no role or should be neglected. As in any "complex disease" (a disease whose predisposition is caused by a large number of genes, each having a limited individual effect), such as type 2 diabetes or high blood pressure, the phenotype (the disease) is the result of the interaction between the genetic background and the environment. Ignoring (often deliberately) this dual influence hinders all therapeutic progress. In this case, we are dealing with genuine "medical obscurantism" (see Part II).

Morin (2025) also highlights this fallacious and manichaean dichotomy and defends the highly questionable idea that priority should go to environmental research, with genetics (in this case, "genome-wide association studies") being accused of providing only doubtful results.

Bindon (2020), Cerdeña, Grubbs and Non (2022), Clerget-Darpoux and Denin (2023), Duncan et al. (2024), Oza (2024), and Peifer (2025) put forward the same manichaean strategy of accusing genetic studies of minimizing the influence of the environment, a strategy also used in the Wikipedia chapter on human races (Anonymous, 2022a).

These accusations are unfounded. No geneticist advocates the idea that the environment plays no role, or a very minor role, in the genesis of phenotypes. The proponents of the "all genetics" view, if they exist, are conspicuously absent, to my knowledge, in the scientific world. One would have to seek them out in the worst conspiracy networks. It is notable that genome-wide association studies now integrate environmental risk factors into their protocols (Southam and Zeggini, 2025), thus refuting the manichaeism denounced by some. Powerful tools, advocating a multigenic approach, taking into account environmental factors, and relying on controlled artificial intelligence, are now being developed, such as PIGEON (Polygenic gene-environment interaction) (Miao, 2025).

The Consensus is Fading

Over the last twenty years, it has become clear that the "consensus" on the non-existence of human races was quite relative and is increasingly eroding, as emphasized by Taguieff (2018).

The consensus on the non-existence of human races is limited to the "Western world," within which it is far from absolute.

In fact, this "consensus" prevails primarily in the "Western world," North America and Western Europe, where it is far from absolute. What is consensus, as we have already seen (see: "other arguments of scientific antiracism"), is the increasing avoidance of the term "race" used in isolation in scientific publications (Byeon et al., 2021). Outside the "Western world," the "consensus" is absolutely not shared by scientific communities in other regions, such as Eastern Europe, Russia, and Asia (Anonymous, 2021a; Štrkalj, 2007), proof if ever there was that a debate exists.

According to Bindon (2020), at the time of the advent of the Human Genome Project in 2000, the race (-) consensus had been strengthened by the fact that the results of this project confirmed that humans were 99.9% genetically identical to each other. But according to this author, during the 21st century, this consensus has been waning.

Lala and Feldman (2024) note that the "consensus" on the non-existence of human races could be viewed with suspicion and seen as having *political* motivations (my emphasis). The risk of this is undeniable.

The Phenotype Speaks: Skulls Tell the Tale

Craniometry, or the study of skulls by measurements, has a bad reputation. It was extensively used and instrumentalized in the 19th century by sexist and racist science. The French anatomist Paul Broca drew "irrefutable" proof of man's superiority over woman from it. The American anthropologist Samuel George Morton "demonstrated" the superiority of Europeans over Africans. We can note here that, while the conclusions of these two scientists have been refuted, their observations have not. The cranial measurements performed by Morton (1849), which had been questioned by the paleontologist Stephen J. Gould in his famous work, "The Mismeasure of Man" (1996), were validated by Lewis et al. (2011).

In opposition to Morton's racist aims, as we saw earlier (see: "other champions of scientific antiracism"), Franz Boas, a leader of American antiracist scientists in the first half of the 20th century, undertook to demonstrate through craniometric studies that cranial morphology was primarily modulated by the environment, and therefore had limited anthropological interest (Boas, 1912). This was refuted by Sparks and Jantz (2002), who conclude that the heritability of cranial morphology is high, even though the environment also plays a notable role. We must emphasize that heritability (see glossary) does not concern the part of a given trait (here, cranial morphology) determined by genetics, as is often believed, but rather, the part of the *variability* of that trait that is of genetic origin.

> *Cranial morphology exhibits notable heritability and is related to the geographical origin of populations.*

More recent studies fully confirm that craniometric analysis can identify, not the level of cognition, but rather, the geographical origin of individuals with great reliability (Ousley et al., 2009, Relethford, 2009). This is even acknowledged by Sauer (1992) and Wade (2021a) who, however, do not accept the biological concept of human races. Wade (2021b) notes that some anthropologists using craniometry prefer to substitute finer identifications (Japanese instead of Asian, for example) for the main geographical categories (Africans, Europeans, etc.). This reinforces the idea that cranial morphology can identify the geographical origin of individuals. Wade (2021a) points out that a trait previously believed to be specifically Neanderthal, namely the morphology of the nasal cavity, is absent in individuals of European descent but is found in some non-Europeans. This is therefore another cranial characteristic that allows for the identification of the geographical origin of individuals.

For cranial morphology, geographical populations exhibit average differences, without a sharp break, with some overlap, thus mimicking the variability of common genetic markers. Ousley et al. (2009), who, incidentally, do not accept the biological concept of race, observe that geographical identification becomes more and more accurate as more craniometric characteristics are added. This indicates that, like genetic markers, these cranial characteristics are not independent, but *congruent*: they all point in the same direction.

Skin Pigmentation

Skin color is obviously the first thing that comes to mind when discussing human biological diversity, given that this characteristic has been used since ancient times to describe human geographical populations. Furthermore, the use of this trait has been heavily caricatured by "scientific antiracists": no one has ever said that the description of human races was limited to noting skin color ("skin deep"). As emphasized by Ayala and Kiger (1985), the description of subspecies or geographical races must be based on all their characteristics, not on isolated traits.

The description of human races has never relied solely on the trait of skin pigmentation.

Moreover, classical anthropologists (Vallois, 1976) have insisted that skin color varies greatly within each major geographical population. Even in highly racist writings of the 20th century (Joubin and Robin, 1923), this was clearly noted. Nevertheless, the distribution of skin pigmentation is clearly linked to geography. Generally speaking, with exceptions, the further from the Equator, the fairer the skin; the closer one gets, the more pigmented it is (Takezawa, 2023). The people with the darkest skin are in Sahelian Africa.

Skin pigmentation appears to be a factor of climatic adaptation (degree of sun exposure). However, some authors have inferred a role for sexual selection in the genesis of this trait.

Recall that the *F*st statistic (see Box 1) measures the genetic differences between populations, between the minimum values of 0 (no differentiation) and 1 (maximum differentiation). For common genetic markers, as we have seen, this value oscillates between 0.05 and 0.15 between main human geographical populations. If we adapt this statistic to skin color, we obtain values between 0.6 and 0.9 (Harpending, 2002; Relethford, 2009), which indicates—as we already knew, but with the benefit of an objective measure—that geographical populations differ considerably in their epidermal pigmentation. However, these differences are not sharp: there are no abrupt discontinuities, which is easy to observe when traveling around the globe.

In my adopted country, Thailand, 90% of the population belongs to the Thai ethnic group (Thailand = Land of the Thais). This ethnic group is considered genetically quite homogeneous. However, Thais show a very wide range of skin colors.

In Thailand, "white is beautiful" (especially for ladies).

Some Thais have very fair skin, which is the ideal of feminine beauty. Moving towards the Cambodian border, people can have very pigmented skin. Finally, some inhabitants of the northeast of the country can have almost black skin.

The absence of abrupt discontinuities in skin color is easily explained by the fact that skin pigmentation is considered primarily the result of climatic adaptation (sun exposure). Now, the climate shows few sudden changes over a few kilometers.

Countries with a highly mixed population like Brazil can show very surprising cases of disjunction between phenotypic traits and genetic substrate. One samba singer, identified as "black" (negro) by everyone, including himself, actually had 67% European genetic ancestry (Rodriguez Mega, 2021).

According to Harpending (2002), sexual selection plays a role in skin color differences. In Thailand, skin color is a determining factor in sexual attraction. Female TV presenters and movie stars are generally selected for their very fair skin. Many of them are actually Eurasian. Women in this country spend fortunes on dangerous and ineffective whitening cosmetics and constantly walk with a parasol to protect themselves from tanning, including while riding their motorcycle. Indeed, very dark skin betrays the peasant woman who spends her days in the sun in her rice field. Having very fair skin, conversely, signifies being a glamorous "city girl." We had the same trend in Europe in the *Belle Époque*, whereas now, there is a rush for tanning, leading to a proliferation of cutaneous melanomas.

Other Phenotypic Traits

Apart from skin color, numerous phenotypic traits, primarily morphological ones, allow for the identification of the geographical origin of individuals with good reliability. Even in a small country like France, the population is not homogeneous. If limited to French people of European descent, faces and statures differ statistically between Dunkirk, close to the Belgian border (many people of Flemish descent) and Montpellier in southern France (population of Mediterranean origin). But these "rough-and-ready" approaches must not replace objective scientific data.

According to Edgar (2009), Western and Eastern Europeans can be distinguished by dental morphology. The differences are even stronger between Americans of European and African descent. By comparing individuals from the current era with people from the 17th century, this author postulated that mixing in the United States had been much more intense between Eastern and Western Europeans than between Americans of European and African descent.

Geographical populations differ considerably in many morphological traits.

Sarich and Miele (2004) postulate that the facial and cranial morphological differences between major geographical populations (e.g., between Asians and Africans) are greater than those observed between bonobos and chimpanzees, which nevertheless belong to different ape species. Furthermore, these differences are 10 times greater than those found between individuals of different sexes within the same geographical population. Their book, "Race: The Reality of Human Differences," has been heavily criticized. It has the drawback of not providing precise bibliographic references. However, it is a goldmine of scientific information. Moreover, the first author, Professor Vince Sarich, is a leading scientist, a pioneer in the study of molecular evolution. His writings therefore deserve to be taken seriously.

According to Gouveia et al. (2025), height and body mass index are correlated with ancestry groups: "White," "Black," "Asian," and others. Within the "Black" group, East and West Africans do not have the same association with body mass index. This confirms that the main geographical groups are genetically and phenotypically heterogeneous, and that additional subdivisions are often necessary for the study of certain characteristics (see Chapter 2: geographical populations and medicine).

> *Like genetic polymorphisms, phenotypic traits are correlated with each other, making it possible to reliably identify the geographical origin of individuals.*

To conclude, the objective statistical analysis of phenotypic characteristics, particularly eye and skin color, facial and cranial morphology, and dentition, makes it possible to reliably identify the geographical origin of any individual. This stems from the fact that these traits are *correlated* (congruence): they are not independent of each other, except, sometimes, in cases of recent mixing.

Blue eyes generally go with fair skin and tend to be more frequent in blonds. Frizzy hair is *statistically* associated with dark skin, etc. We have already seen this phenomenon of congruence with genetic characteristics. Mixing can sometimes break these statistical associations. We already mentioned the case of the Brazilian samba singer identified as "black," despite having 67% European genetic ancestry. I remember a farmer in the Bolivian Altiplano who had the typical physiognomy of an Aymara Indian, except that his almond-shaped eyes were an unusual azure blue. However, as we have seen (Parra et al., 2004), a very significant correlation is observed between skin color (objectively measured by spectrophotometer) and ancestry (assessed by informative ancestry markers). This disjunction of characteristics in mixed-race individuals is therefore not the rule.

We have seen that the concept of phenotype is not limited to morphological traits but includes all characteristics, including physiological ones, apart from the genome. Particular electrocardiographic features (repolarization abnormalities) are observed in a certain proportion of individuals of African descent; in most cases, these have no pathological significance, but they could mislead a physician unaware of this specificity (Gentillini et al., 2012).

A curiously neglected approach in the study of human geographical populations would allow for a better assessment of this phenotypic diversity: it is *numerical taxonomy* (Sneath and Sokal, 1973). This approach consists of taking into account as many phenotypic traits as possible and evaluating the resemblance between taxa by computer processing of this data. No hierarchy between phenotypic characteristics is established *a priori*. This approach has proven very effective for the classification of many species, but has never been applied, to my knowledge, to human geographical populations. It could be effectively brought up to date with the contribution of big data computing and artificial intelligence.

Diseases and cognition are also phenotypic traits. However, they fully warrant specific developments (see Chapter 2: geographical populations and medicine and Chapter 3: recent evolution of cognitive traits).

Breeds in Dogs, but No Races in Humans?

According to Jordan (2008), the genetic variability of man's best friend, *Canis lupus familiaris*, is comparable to that of humans, but it is primarily observed *between* dog breeds, with each breed being genetically very homogeneous. This would show, according to this author, that the concept of race is acceptable for dogs, but not for humans.

> *The differences between dog breeds and between human geographical populations are generated by similar genetic mechanisms. The magnitude of these differences is a question of degree, not of nature.*

In fact, the Figures given by vonHoldt et al. (2010) do not support this. They report a genetic variability in dogs of 65% within breeds and 35% between breeds. Cochran and Harpending (2009) give similar results: 70% and 30%. Recall that in humans, 85% to 95% of genetic variability is observed within continental populations, and 15% to 5% between them, if one limits oneself to only "common" markers. Between dogs and humans, there is therefore a difference of degree, and not of nature, regarding intra-breed (intra-continental population) and inter-breed (inter-continental population) variability.

The Highs and Lows of Artificial Intelligence

The progress of artificial intelligence (AI), particularly in the field of medical imaging, is hampered by the same problem as for genomic studies: the data have been primarily generated based on populations of European ancestry, which makes them less applicable to other populations (Rao et al., 2025).

Despite this limitation, artificial intelligence can already yield surprising results regarding the differentiation of geographical populations.

> *Artificial Intelligence appears capable of identifying a patient's geographical origin from their mammogram, which a human observer cannot do.*

Thus, through mammographic analysis, AI would be able to determine a patient's origin: European, African or African American, Asian or "Latino", even with poor quality images (Adleberg et al., 2022; Gichoya et al., 2022). No gynecologist is capable of this. The fact that artificial intelligence can also deduce a patient's geographical origin from chest X-rays or cardiac ultrasounds, when no human operator is capable of it, is considered "puzzling" by Zou et al. (2023).

However, artificial intelligence, for now at least, has human creators, and tends to reproduce what has been put into it. It has been accused of faithfully reflecting the "politically correct" ideology of its Silicon Valley creators (Mounk, 2024). If one is curious enough to ask them the question, "Do human races exist?," one is struck by the ultra-orthodox answers from Google Gemini and ChatGPT, a pure distillate of all the classic arguments of "scientific antiracism": no sharp separation between geographical groups, the *F*st argument, genetic continuum. Of course, there is no mention of the caveat introduced into the *F*st by low-frequency or rare genetic markers. Carried away by their "politically correct" momentum, artificial intelligences (Google Gemini) generate oxymoronic images of African-type Waffen SS and Kings of England (Mounk, 2024).

In contrast to these antiracist tendencies, AI softwares have been accused of spreading racist prejudices. For example, AI has been criticized for reproducing racist stereotypes when confronted with texts written in the slang favored by African Americans (Blodgett and Talat, 2024; Hofmann et al., 2024).

Some of these software programs would even continue to spread false medical beliefs (Omiye et al., 2023). In fact, the examples cited in this work are erroneous, or at the very least highly debatable. The most illustrative case is the authors› assertion that lung capacity does not vary between geographical groups. I will elaborate later (see Chapter 3: altitude adaptation) on the case of the Aymara Indians, inhabitants of the Bolivian and Peruvian Altiplano, who show remarkable adaptation to altitude, particularly increased lung capacity. Other geographical populations exhibit variations in lung capacity, variations affected by the degree of miscegenation. These differences impact the frequency of asthma (Mersha and Abebe, 2015; Kumar et al., 2010). The authors' claim that differences in renal function between geographical groups "have been refuted" (Omiye et al., 2023) is also false. If geographical origin in the exploration of renal function is not taken into account, it leads to an underestimation of the glomerular filtration rate in African Americans (Hsu et al., 2021). Similarly, contrary to the authors› statements, there indeed seems to be differences in skin sensitivity between geographical groups (Kim et al., 2017). The same applies to the affirmation of the absence of skin thickness difference between Americans of African and European descent. Such differences, although qualified as "minor," do exist (Table 4 in Lintzeri et al., 2022). Fighting racism is a noble cause. Imagining that one will Fight it effectively by stating factually false or debatable arguments in a scientific journal is a serious mistake that harms this noble cause (see Part II, Chapter 3: The Threat of Scientific and Medical Obscurantism).

Finally, a recent application of artificial intelligence has been to contribute to an "equitable" genomic precision medicine through the machine learning method PhyloFrame (Smith et al., 2025). This approach aims to remedy the frequently denounced bias of genomic samples where Europeans are overrepresented. The method corrects this "ancestry bias" by integrating population genomics data with transcriptomic* data. It appears more satisfactory than self-declared ancestry because it allows, for example, better consideration of the considerable genetic diversity observed in individuals of African descent, instead of considering them as a single category: "Africans."

In summary: artificial intelligence does not always go where its creators would like it to go. However, potentially, if its current imperfections are corrected, it will certainly enable considerable scientific advances, unless it enslaves us in the meantime! (Alexandre, 2024).

The "Genetic Discovery Bias": Too Many "Whites"?

> *In most biomedical studies, individuals of European ancestry are overrepresented, which leads to missing a very important part of the genetic diversity of our species.*

It is repeatedly stated in these studies that populations of European origin are overrepresented, which introduces a considerable bias in the analyses. Indeed, European populations are among those that exhibit the least genetic and genomic variability. By focusing excessively on them, a considerable part of the overall genetic variability of the human species is overlooked. This is what is called the "genetic discovery bias" (Bien et al., 2019; Gouveia et al., 2025; Kullo, 2024; Murray, 2020; Reardon, 2023; Thomas and Peters, 2025; Wang et al., 2024). Ju et al. (2022) communicate eloquent Figures: of all the genome-wide association studies surveyed by them, 67% concerned Europeans, 17% East Asians, 7.5% African Americans or Afro-Caribbeans, and 3.8% Latin Americans or Hispanics. Africans from Africa constituted an extremely small percentage. Examples of large differences in frequencies (up to 10 times more) of traits involved in pathologies across different geographical populations are numerous. These traits would have been missed if sampling had been limited to populations of European origin (Karczewski et al., 2025).

The contribution of Oni-Orisan et al. (2021) regarding this problem deserves to be highlighted. Indeed, the authors of this article are all of African ancestry and are therefore unlikely to be suspected of "systemic racism." They are seasoned experts, with strong scientific or medical backgrounds. They insist that considering the entirety of human genetic diversity is necessary to improve "black health." Finally, they move away from the refrain "race is a purely sociological construct" with this almost iconoclastic statement: "we believe that race has both a genetic and a social component." The authors are nonetheless convinced and declared antiracists. According to them, social science researchers and geneticists must collaborate to develop antiracist but not anti-genetic curricula. Their conviction is that genetic research, if responsible, is antiracist. I wholeheartedly share it.

The Origin of Human Races

> ***The cradle of humankind is a cradle on wheels***
> ***(Father Henri Breuil)***

The "Out-of-Africa" (OOA; Stringer and Andrews, 1988)

This is the current dogma, at least among Western scientists: all modern humans have populations from the African continent as ancestors. This is where the "cradle of humanity" is located. However, I always take mischievous pleasure in citing the very witty remark of the famous French prehistorian Father Henri Breuil (above), because the least that can be said is that the cradle is restless.

*A majority of scientists place the origin of modern humans (*Homo sapiens*) in Africa. However, the dates of this "Out-of-Africa" and the modalities of genetic exchange with local archaic populations are debated.*

In many people's minds, the Out-of-Africa concerns only modern humans. However, it is probable that there were several successive exits. *Homo erectus* (the former "Pithecanthropus") would have attempted the first trek about 1.8 to 1.9 million years ago (Michel, 2018; Templeton, 2017). The Dmanisi fossils in Georgia, 1.8 million years old, attest to the antiquity of the exit from Africa (Bettayeb, 2025). The discovery of the Longgupo jaw in China could indicate the presence of *H. erectus* in this country as early as 1.8 million years ago (Michel, 2018), which would imply that the Out-of-Africa of *H. erectus* would be older than supposed. Tools discovered at Shangchen (China) have been dated to 2.2 million years ago. The first exits from Africa thus seem very ancient.

A second exit from Africa is thought to have occurred about 650,000 years ago, in the Middle Pleistocene*.

The last exit from Africa is that of those called modern humans, meaning those who would pass unnoticed in the subway if they wore modern clothes. According to Templeton (2017), the second and third waves resulted in interbreeding with the Eurasian populations of "archaic humans" already present. This already dilutes the "pure and hard" OOA a bit.

A strong argument in favor of the Out-of-Africa hypothesis is that current Africans carry the highest rate of genetic polymorphism among current human populations. Other geographical groups appear as portions, or subsets, of the African genetic variability. This is a classic population genetics argument for considering African populations as original.

Former Models of Human Population Origins

The famous anthropologist Carlton S. Coon proposed the "Candelabra" model (1962). According to this author, the main current geographical groups (Europeans, Africans, Asians) originated from distinct *H. erectus* populations, which would have evolved separately from one another. This hypothesis does not fit with population genetics data, which clearly indicate the persistence of gene flow between geographical populations. The Multiregional Model of Weidenreich (1946) is also largely incompatible with population genetics data. It postulated that all human populations since *H. erectus* constituted a single species, but with regional evolutions for certain morphological traits.

Which Date for the Out-of-Africa?

There is no consensus on this subject. For the exit of modern humans, the consensus was long considered to be 50,000 years ago (Fumagalli and Balloux, 2017; Hublin, 2011). Cavalli-Sforza and Feldman (2003) propose an exit date of 100,000 years ago. Templeton (2017) sets the bar as high as 130,000 years ago.

Discoveries in Southeast Asia suggest the existence of "pre-OOA." Indeed, *Homo sapiens* fossils significantly older than 50,000 years have been discovered in China (80,000–120,000 years), Laos (68,000–86,000 years), and the Philippines (67,000 years). These *sapiens* may not have left genetic traces in current populations, although this is debated (Martinón-Torres and Lalueza-Fox, 2025).

In short: there are still many questions surrounding the OOA.

The Out-of-Africa Model has Shifted

The "pure and hard" model with a single Out-of-Africa and complete replacement of pre-existing local populations is losing ground. Harari (2011) insinuated that it was tinged with "political correctness," as it insisted on the genetic homogeneity of humanity. This model has been qualified as "simplistic" by Vasseur and Quintana-Murci (2013).

It is now generally accepted that modern humans, while spreading across the globe, interbred with local populations, the "archaics," Neanderthals and Denisovans*, who had themselves exchanged genes with each other (Alves et al., 2012; Pääbo, 2015; Templeton, 2017). This has been called "adaptive archaic introgression"* (AAI*), because our archaic cousins, in welcoming the "moderns," offered them genes that allowed them to "skip steps" to adapt more quickly to the climates of the host countries and to local infectious diseases. This "genetic hitchhiking" contributed to the genetic differentiation of humanity, since only non-African populations benefited from it, and unevenly so. The hitchhiking was, moreover, an exchange of good deeds, since the "moderns" offered some of their genes to the "archaics" (Gibbons, 2017a; Kuhlwilm et al., 2016).

Furthermore, the "classic" Out-of-Africa model considered the east of this continent as the unique "cradle of humankind"; many now believe in a multi-regional origin of *Homo sapiens* in Africa, not limited to East Africa.

We therefore have a kind of revisited multi-regional model, with advanced local adaptations, favored by interbreeding with the archaics who were already present. Apart from certain specificities of human evolution and migrations, particularly the fact that our species populated all continents, covering a wide range of climates and ecosystems, this picture evokes a classical process of the appearance of subspecies (or geographical races), through allopatry*, differential local adaptations, natural selection in response to different environments, and genetic drift*.

The reanalysis of the "Yunxian 2" skull (China), one million years old, which had been attributed to *H. erectus*, could fundamentally change our view of the OOA. Indeed, it might rather be attributable to *H. longi*, or even *H. sapiens* (Birken, 2025; Feng et al., 2025). This could suggest a major phylogenetic subdivision one million years ago in Asia. This continent would then appear as the cradle of humanity, fundamentally questioning the OOA. Thus, our hypotheses, based on a limited number of fossils, often in poor condition, appear quite fragile. More than ever, to use the words of Father Breuil, "the cradle of humankind is a cradle on wheels."

How to Classify Human Geographical Populations?

Why classify them, by the way? Because this taxonomic approach is essential in all science, particularly in biology. To study the properties of living beings, it is essential to classify them, to identify the units of analysis that will be compared with each other. This inventory work comes upstream of everything else. It is not there to satisfy a collector's mania, but on the contrary, it has considerable practical implications.

> *Classifying living beings (taxonomy) is an indispensable step to describe their properties. This is also true for human geographical populations.*

If a microbiologist endeavors to differentiate *Staphylococcus aureus* from *Staphylococcus epidermidis*, it is primarily because the former is generally much more pathogenic: identifying it has strong predictive power. Therefore, classifications are indispensable. But to be reliable and operational, they must meet two essential criteria. First, the criteria on which they are based must be clearly stated. Then, their *predictive power* must be evaluated, because this is one of the major interests of any classification. The identification of a unit of analysis, of a *taxon*, is a *working hypothesis*, a model. Its predictive power, if verified, reinforces it. Otherwise, it discredits it.

A Model with Low Predictive Power: "Move Along, Nothing to See Here"

I find this slightly provocative statement very illustrative of the doxa "human races have no biological basis," which suggests more or less implicitly that human geographical variability is a phenomenon only of interest from a sociological point of view, by generating condemnable behaviors such as racism, slavery, and discrimination of all kinds. This sociological aspect is indisputable: differences between human populations have been extensively exploited for these criminal purposes.

> *The doxa "human races do not exist" suggests that the biological differences between geographical populations are negligible, meaning they deserve to be ignored, and are moreover in the process of being so, including by medicine ("medical obscurantism").*

But what this doxa implicitly suggests is that on the physical, biological, and genetic level, the geographical variability of our species would be negligible, trivial, and unworthy of interest. Differences would be limited primarily to skin color or nose shape (Foucart, 2025). According to this model, human races, for all the reasons explained above ("the debate: scientific antiracism"), are a vague concept, almost meaningless from a biological point of view. All these classical arguments are perfectly reproduced in the presentation of the question by Google Gemini and ChatGPT.

Thus, if one follows the predictions of this model, if one compares, for example, Congolese with Chinese or Australian Aborigines, one will not find much, apart from different skin colors and a few facial features. This is simply false. We have already seen regarding genetic variability that the famous "*F*st argument" is flawed if one considers low-frequency or rare genetic variants, variants that play a large role in the genesis of phenotypes, especially diseases. These variants, in fact, tend to have very different frequencies from one geographical population to another. Furthermore, if one followed the predictions of the "nothing to see here" model, by scrutinizing the genetic diversity of human populations, there should be a total discordance between this genetic diversity and the major geographical groups. However, the opposite is obviously true (Figures 4–8).

I have already presented the case of considerable morphological differences between geographical populations, differences that exceed those that can be observed between two different species of primates, the common chimpanzee and the bonobo (Sarich and Miele, 2004). But it is on the medical level that the "nothing to see here" model is least satisfactory. I will explain in detail (Chapter 2: geographical populations and medicine) why it is crucial to take into account human geographical variability in biomedical research and medicine, and why (part II, Chapter 3) "scientific and medical obscurantism" is a growing threat.

If one wants to go beyond the "nothing to see here" model and bring order to the dense forest of human biological variability, a priori, one embarks on a path full of pitfalls. I have extensively discussed the continuum argument, constantly put forward by defenders of scientific antiracism. But no entity can claim perfection, except in mathematics and philosophy. "Nature laughs at our classifications," as I have already quoted. However, since Linnaeus, despite the snickering of Mother Nature, naturalists have strived to classify living beings, and indeed, they have bequeathed to us models that, quite often, hold up perfectly, despite not being perfect, "essential" entities (provided with a specific "essence"). It is remarkable that an enormous number of species described by our great predecessors (Linnaeus, Buffon, Daubenton, Lamarck) have proven to be "good" species in the light of modern genetics. As for human geographical populations, what have our predecessors bequeathed to us?

Antiquity of the Concept of Race

A false idea that is frequently encountered (see above: "history of the notions of race and racism"), is that race is a modern invention, dating in fact from the important scientific progress of the Renaissance, and that it was forged just in time to justify colonialism and slavery. There is no doubt that these two harmful ideologies have used and abused the interiorization of people of color to better enslave them.

Peoples of antiquity distinguished races.

However, ancient peoples had eyes to see and minds to think. According to Sarich and Miele (2004), Greeks, Romans, Egyptians, Indians, Chinese, and Arabs distinguished races, often attributed them to climate, and considered them hereditary.

Phenotypic differences between geographical populations are attested by numerous artworks dating back to antiquity.

Classifications of the Modern Era (Since the Renaissance)

From the Renaissance to the present day, scholars have established racial classifications that differ from one another, but nonetheless distinguish the main geographical populations. Some of these classifications include cognitive and behavioral traits in their descriptions and establish a hierarchy *between races.*

As already exposed (see "do human races exist? The list them"), Linnaeus, the father of modern systematics, in the 2nd edition of his major work *Systema Natura* (1740), proposed the following classification: "Europaeus albus" (white Europeans); "Americanus rubescens" (red-skinned, Native Americans); "Asiaticus fuscus" (yellow Asians); "Africanus niger" (black Africans). He added a rather fanciful category, "monstruosus". He attributed not only physical but also psychic traits to these categories, sowing the seeds of racism in its most complete form (Anonymous, 2022a).

The German physician and anthropologist Johann Friedrich Blumenbach adopted the same classification in 1795, and added a 5th race, the "Malay" (Anonymous, 2022a).

The French naturalist Georges-Louis Leclerc, Count of Buffon (1707–1788), distinguished 6 geographical varieties. He was an adherent of the "degeneration" theory (Patou-Mathis, 2013), according to which, originally, all humans were white, the other skin colors having been generated by degradation, by moving away from the temperate zone. Buffon's varieties were: Europeus (white, sanguine, energetic); americanus (red, choleric, upright); asiaticus (yellow, melancholic and rigid), afer (black, phlegmatic, relaxed). In Buffon, one finds the mixture of physical and psychic traits already present in Linnaeus. The theory of degeneration was adopted by Blumenbach.

The famous French paleontologist George Cuvier (1769–1832) distinguished 3 races (Mohsen, 2020): Caucasians (= Europeans), Mongolians (Asians), Ethiopians (Africans). He considered Europeans "more beautiful," and Africans, "ape-like," subjective judgments if ever there were any.

The German biologist and philosopher Ernst Haeckel (1834–1919) distinguished 10 races and considered Caucasians (Europeans) as "superior" (Anonymous, 2022a).

The American Theodore Lothrop Stoddard distinguished "Nordic," "Alpine," and "Mediterranean" races in Europe.

More recently, Theodore Lothrop Stoddard, an American historian and journalist (1883–1950) distinguished, within Europeans, "Nordic," "Alpine," and "Mediterranean" races. He considered the Nordics as "superior" (Anonymous,

2022b). This classification had a certain longevity, since it still appeared in my High-school geography class in the 1960s.

> *The American Coon introduced the concept of population genetics into the definition of human races.*

The American anthropologist Carleton S. Coon (1904–1981) distinguished 5 main races (Anonymous, 2022b): Caucasoids (Europeans), Mongoloids (Asians, including Amerindians), Australoids (Australian Aborigines), Congoids (Africans), Capoids (Hottentots) (Hebert, 1977). Coon introduced a major innovation by considering human races as evolving Mendelian populations*, thus giving them a basis in population genetics (Dobzhansky, 1962a).

The French anthropologist Henri Vallois, author of the famous book "les races humaines" (Human Races) (1976), considered human races as subspecies in the famous treatise on zoology by Pierre-Paul Grassé, In accordance with Linnaean terminology, he assigned them trinames: *Homo sapiens australosicus* (Australian Aborigines), *H. s. albus* (Europeans), *H. s. asiaticus* (Asians), *H. s. afer* (Africans) (Dobzhansky, 1964). These subspecies were what he called the primary or great races (Hebert, 1977).

The American anthropologist Stanley Marion Garn (1922–2007) distinguished the following groups: (1) American Indians; (2) Polynesians; (3) Micronesians; (4) Papuans-Melanesians; (5) Australians; (6) Asians; (7) Indians; (8) Europeans (including North Africa and the Near East); (9) Africans (Dobzhansky, 1964).

Egon von Eickstedt (1892–1965), a German anthropologist, distinguished 3 races, europiforme or leucoderme (white), negroforme or melanodermic (black), and mongoliforme or xanthodermic (yellow) with 29 sub-races (Dobzhansky, 1964).

The Swedish anthropologist Bertil Lundman (1899–1993) described 4 main races in 1967: white, yellow, red, black, as well as 16 sub-races (Dobzhansky, 1973).

In the United States, where ethnic statistics are routine, unlike in France, it is very common to be questioned about one's ethnic origin for many administrative procedures (although this declaration is optional). The last census in 1977 proposed the following categories (Bindon, 2020): (1) Native American, including Alaska Natives; (2) Asian, including Pacific Islanders; (3) Black; (4) White; (5) Hispanic.

Finally, the latest edition of the Dictionary of Genetics (King, Mulligan and Stanfield, 2012) describes Africans, Asians, Caucasians (Europeans, including North Africa, India, and the Middle East), Pacific Islanders (including Australian Aborigines and New Guinea residents), and finally, Native Americans.

The "Major Geographical Groups": A Very Resilient Classification

From this rather tedious enumeration, one might feel an impression of inconsistency, of arbitrariness. One author, like Carleton S. Coon, successively described 9 races, then 32. Another author sets a record with 200 races! (Dobzhansky, 1962a). These apparent divergences have been taken as an argument to attack the concept of biological race (Foucart, 2013). However, the inconsistencies are only apparent.

There is in fact a lot of overlap between the different classifications. The three classic "great races" (Europeans, Africans, Asians) consistently emerge. The categories described in the recent work that is the Dictionary of Genetics (King et al., 2012), namely Africans, Asians, Caucasians (= Europeans), Pacific Islanders, and Amerindians, are not very far from those identified by Linnaeus and Blumenbach. The same applies to the categories proposed by the "Genome Wide Association Studies diversity monitor": Europeans, Asians, Africans, African Americans or Afro-Caribbeans, Hispanics or Latin Americans, other/mixed (Corpas et al., 2025). In this classification, the last two categories are of course unsatisfactory catch-all terms from a biological perspective, as emphasized by Corpas et al. (2025). However, once again, the main continental groups emerge.

The main geographical groups persist from one classification to another.

Any classification inherently contains an element of arbitrariness and cannot aim for perfection. As stated above (see: "Human races are difficult or impossible to define from a biological point of view"), autoimmune diseases and, even more so, mental disorders, are based on the description of categories full of problems. In comparison, the major geographical groups appear as models with satisfactory resilience and predictive power.

As stated above (see: "Do human races exist? The list them!"), according to Dobzhansky (1962a), the existence of biological races and the differences between them is a biological fact, while their enumeration is a problem of nomenclature, exactly like developing a classification system in a library: classical/modern authors, American/foreign authors, specialized/popular science works. Everything depends on the goal sought. What is certain is that the admission of failure that is the "Nothing to see here" model is little followed by researchers in the biomedical field.

The doxa "human races have no biological existence" is little followed by researchers in the biomedical field.

Indeed, a considerable number of publications take the major geographical groups as the basic sample, even if many authors (not all: see Figure 7 and: (https://drive.google.com/file/d/1qWNbPeyBCwDL9CgzoOMrHguFm9MeX4f2/view). (https://drive.google.com/file/d/1pffskqiM3WuPeD7gLIWYsSiXyTdwoeso/view) avoid the term "race" in isolation, or not accompanied by diplomatic quotation marks.

Chapter 2

Geographical Populations and Medicine

The study of human geographical populations raises fascinating philosophical, historical, and political questions. However, it has a very utilitarian interest, which is medicine: geographical populations are differentiated with respect to their susceptibility to a number of diseases, and these differences are in part innate, or genetic. Taking this fact into account, which is well-known to specialists in tropical medicine (Gentillini, 1986, 2012), is essential for developing personalized medicine (Lee and Mudaliar, 2009).

Differences appear much more clearly at the population level than at the individual scale. At the individual level, exceptions that confirm the rule will always be found. The fact that some tall pygmies surpass some short Swedes in height should not lead to the conclusion that these two populations do not differ in the "height" phenotype. A population-based approach is relevant for *preventive medicine*, which concerns pathological phenomena of statistical interest to populations.

> *It is crucial to take human geographical diversity into account in the fields of personalized medicine, preventive medicine, and public health.*

This unequal distribution of certain diseases is entirely understandable. Since the migration out of the African continent, human populations, by dispersing across the globe, have been subjected to distinct selective pressures, varied climatic conditions, as well as different pathogens. This phenomenon still persists today. Diseases, like any other phenotype, result from the interaction between the environment and the genetic makeup of individuals and populations.

DOI: 10.1201/9781003647393-3

Different Strategies for Analyzing the Genetic Control of Diseases: Candidate Genes, Genome-Wide Association Studies, Whole Genome Sequencing, Common Disease/Common Variants, Low-Frequency and Rare Variants

Different approaches can be implemented to explore the genetic basis of human susceptibility to diseases. The most classical, but not the most effective, and quite outdated, involves searching for "candidate genes," i.e., genes presumed to play an important role in the onset of diseases. This presumption is based on the fact that their association with diseases has long been known, such as the genes of the HLA system* in the case of communicable diseases or autoimmune conditions. Alternatively, their links with pathological traits have been revealed through animal experimentation. However, caution is needed with this latter approach. A gene does not necessarily have the same role in mice and humans just because it has the same DNA sequence in both species. Mother Nature is a great tinkerer. Exaptation* is the phenomenon by which a given trait acquires functions different from those it originally had. This is a well-known mode of evolution. An illustrative case is the flippers of dolphins, which were legs back when the ancestors of dolphins roamed on land. What is observed in a phenotypic character like dolphin flippers can also be observed at the gene level. The harvest reaped by candidate gene research has not lived up to expectations.

In "association studies" (linkage approach), a given segment of the genome is tagged by genetic markers such as SNPs or microsatellites. A statistical link between the presence of this genomic segment and a given pathological trait is then evaluated in pedigrees, or families. This is a very laborious screening method. The identified segment can contain hundreds or thousands of genes. Once the link between the segment and the pathological trait is established, it remains to meticulously dissect this segment to identify the responsible gene(s).

The study of the genetic control of diseases has been revolutionized by genomics, with genome-wide association studies. This was made possible by the considerable reduction in the cost of these technologies.

These candidate gene/association study methods tend to be advantageously replaced by global genomic studies, made easier by advances in high-throughput computing and genomics, with a considerable lowering of the cost of these analyses: the first human genome sequencing, released on February 16, 2001, cost over a billion dollars. Nowadays, the company Illumina sequences an entire human genome for about 1000 and is capable of doing 100 sequences per week. Roche, the largest pharmaceutical company in the world, promises even better with its new SBX (sequencing by extension) technology (Jordan, 2025).

An approach now employed on a very large scale, and which tends to become generalized, is the genome-wide association study (or GWAS), which is an association study extended to the entire genome. Its goal is to identify the genomic region(s) where the genes responsible for the studied traits are located, whether or not these traits are pathological. It relies on the complete tagging of the genome with high-resolution markers, by comparing individuals carrying the trait under study with individuals who do not present this trait.

The genome-wide association study (GWAS) tags the entire genome with genetic markers and explores the statistical association of given genomic regions with given diseases, or, more generally, given phenotypic traits ("polygenic risk score"; PRS*).*

This approach makes it possible to develop "polygenic risk scores" (PRS), meaning identifying a series of genes supposed to be statistically linked to a given pathological risk or phenotypic trait.

Despite (or because of?) its effectiveness, which seems to be consensus (Abdellaoui et al., 2023; Southam and Zeggini, 2025), this technology has been harshly criticized. Genome-wide associations have even been accused of having no scientific value (Foucart, 2023; Morin, 2024). In this case, it is up to the critics, who are a clear minority, to provide arguments. It can be noted that highly competent researchers are among the detractors (Herzig et al., 2022). The debate is therefore open. The fact remains that this approach is developing very rapidly.

It is worth noting that the journal *Nature* recently called (Anonymous, 2025) for a concerted effort to reflect on polygenic risk scores, which could potentially be able to screen embryos for the risk of pathologies such as Alzheimer's disease, schizophrenia, severe depressive disorders, type 2 diabetes, and coronary disease. This constitutes an implicit recognition of the effectiveness of these polygenic risk scores.

Another criticism leveled against this technology is that genome-wide associations do not provide causal explanations. Indeed, this is not their goal, which is only to identify the genes statistically associated with the phenotype under study. It is the analysis of these genes that will eventually provide the causal explanation.

Another criticism is that this technology assumes that genes act independently of the environment. No one supports this extreme thesis. I have presented this polemical approach earlier (see: "Erroneous nature/nurture dichotomy"). In fact, the very generally accepted basic notion is that every phenotype results from the interaction between genes and environment. The different criticisms of genome-wide association studies have been reviewed by Foucart (2023).

What are the Most Relevant Units of Analysis for the Study of Human Diseases?

The primary task of any taxonomist facing the organism he wishes to study is to delineate and characterize the "relevant units of analysis" (Tibayrenc and Ayala, 2017a) that will form the basis of all subsequent studies concerning this organism. These relevant units of analysis must: (1) be clearly characterizable; (2) have highly predictive properties. What are these units in the case of the human species?

First, What is the Problem Considered?

The answer to this question depends on the study's purpose. To have a satisfactory idea of the genetic diversity of our species, even a limited diverse geographical sample is better than restricting oneself to a single human geographical group or conducting a blind survey without characterization criteria.

> *For the study of the genetic causes of diseases, one must avoid sampling composed of a single geographical group or sampling without a geographical selection criterion.*

Many biomedical studies, if not most, have for too long been limited to populations of European ancestry. However, Europeans constitute only a small fraction of the total genetic diversity of our species. By limiting oneself to a sample solely composed of Europeans, one misses many genes absent in them but present in other populations. This is called "genetic discovery bias," which I have already discussed (see above: "The 'genetic discovery bias'; too many 'whites'?"). The "HapMap" project (https://www.genome.gov/10001688/international-hapmap-project) and many biomedical programs using the same sampling have been based on three major geographical groups: Africans, Europeans, Asians: (https://drive.google.com/file/d/1pffskqiM3WuPeD7gLIWYsSiXyTdwoeso/view).

> *Many studies are based on the main continental populations: Africans, Europeans, Asians. Some studies extend their sampling to other geographical groups (Oceania, Americas). However, in certain cases, the analysis of populations more restricted than the main continental groups is required.*

For more precise studies, it can be informative to examine much more restricted geographical populations. In Africa, the Fulani ethnic group for the study of malaria is one example. Indeed, the Fulani exhibit a higher resistance to malaria than that of other ethnic groups sharing the same territories (Arama et al., 2015). Along the same lines, Khan et al. (2022) note that the Yoruba ethnic group alone, which is part of the sampling of the international HapMap project, cannot be considered representative of the genetic diversity of all populations of African descent.

Thomas and Peters (2025) cite the example of Alaskan Natives, who are afflicted with a colon cancer rate much higher than that of other geographical populations, including Native Americans. However, Alaskan Natives and Native Americans are genetically very closely related, and they are grouped together in many studies, which prevents seeing what is specific to Alaskans.

Wojcik (2025) cites a much more debatable example, that of Indians (from India) and Chinese, who are classified by the American census system in the same "Asian" category, even though the incidence of thalassemia, a genetic blood disorder, is very different between the two groups. However, no one among biologists, or even in the public, would think of classifying the Chinese and Indians in the same geographical group. This is merely a bureaucratic blunder by the American census system. The other case cited by Wojcik (2025) is more relevant: it is that of Steel syndrome, a genetic bone disorder, which is more frequent in Puerto Rico than in the Dominican Republic or Mexico. Wojcik (2025) thus illustrates the fact that the very broad and highly heterogeneous category of "Hispanics/Latinos" has insufficient predictive power for certain diseases.

The genetic distance between geographical populations can have a strong impact on "polygenic risk scores," and therefore on the possibilities of predicting phenotypes, including at the microgeographical level (e.g., within European populations). This is also true for pathological phenotypes, such as diseases (Privé et al., 2022). The greater the genetic distance between two geographical populations, the less effective the polygenic scores established in one of them will be for the other. In other words, the genetic mechanisms of a phenotype (a disease, for example) highlighted in Europeans will often be of limited use in Africans. This is why it is essential to develop these polygenic scores in several geographical populations.

What about "Races"?

There is little doubt that the main continental groups characterized by genetics quite closely overlap (Rosenberg et al., 2005) with the main human races traditionally identified, the "main races," described by classical physical anthropology (Coon, 1962; Vallois, 1976). Banda et al. (2015) note a strong agreement between "genetic ancestry" and "self-reported ancestry" (SRA) on the one hand, and the social (= socially defined) categories of race/ethnicity on the other.

> *Human races, according to the definitions given by classical physical anthropology and population genetics, are comparable to subspecies or geographical races. However, the term "race" is politically loaded due to tragic historical events (colonialism, slavery, Nazism), which makes its usage problematic.*

Moreover, the genetic diversity observed within and between these continental groups is not inconsistent with the definition of races given by population genetics (Ayala and Kiger, 1984; Dobzhansky, 1962a; Hartl, 1985; King and Stanfield, 1990; King et al., 2012; Mayr, 1963, 2002) and classical physical anthropology (Coon, 1962; Vallois, 1976), which equate these races to subspecies or geographical races (Tibayrenc, 2017b). However, the term "race" itself can be considerably offensive for tragic historical reasons. This delicate semantic problem will be discussed in detail in Part II (ethics and politics). This should not prevent the use of the main geographical groups as units of analysis for the study of diseases, preferably over the "Nothing to see here" model. Moreover, this recommendation is superfluous, because *this is what happens in practice*. The vast majority of biomedical studies take the main geographical groups (Africans, Europeans, Asians, etc.) as study units, even when they avoid using the divisive term "race" (see updated list at: (https://drive.google.com/file/d/1pffskqiM3WuPeD7gLIWYsSiXyTdwoeso/view).

However, the danger comes from the fact that certain medical authorities are now recommending not to take these continental categories into account in diagnosis and treatment (see "Part II: ethics and politics; Chapter 3: The Threat of Scientific and Medical Obscurantism").

Gouveia et al. (2025), for the reason that the genetic groupings of human populations do not 100% match the racial and ethnic categories proposed for the self-reported ancestry protocol, advise against using this approach, without proposing an alternative sampling method. As we have seen, the vast majority of researchers do not follow this recommendation (see above).

Gouveia et al. (2025), while clearly advising against this sampling strategy, contradict their own recommendation by basing their study on the following series: (i) Asians; (ii) Black or African American; (iii) Middle Eastern or North African; (iv) Native Hawaiian or Other Pacific Islander; (v) White; (vi) subjects identified as more than one population; (vii) information not available; (viii) Hispanic/Latino. Such sampling, which matches classical racial and ethnic classifications, certainly provides a very valuable overview of the genetic and geographical diversity of our species, even if, as we have seen, in some studies, finer subdivisions provide valuable complementary information.

To summarize, in any applied study, particularly in the analysis of human diseases, anything is better than a mono-ethnic sampling; for example, Africans and Europeans; even better if Asians are added. The sampling of Gouveia et al. (2025), despite their negative opinion on the self-reported ancestry protocol, has a satisfactory exhaustive nature.

Based on their genetic control, three categories of diseases are distinguished: mitochondrial diseases, Mendelian diseases, and "complex" diseases.

Mitochondrial Diseases

Mitochondrial diseases are a group of genetic disorders that affect the mitochondria, organelles involved in cellular respiration.*

When mitochondria do not function correctly, it can lead to a wide range of health problems. These diseases are generally rare. A list of the main mitochondrial disorders can be found at:
https://www.ninds.nih.gov/search?search_api_fulltext=mitochondrial+disorders (source: National Institutes of Health in the United States). I have not found indications of incidence differences between geographical populations for these mitochondrial diseases per se.

However, certain conditions that cannot be strictly classified as mitochondrial diseases are clearly linked to mitochondrial DNA (mtDNA) polymorphisms. These conditions may be unevenly distributed among geographical populations. Here are a few examples:

Byappanahalli et al. (2024) studied a cohort of African Americans and Americans of European descent identified by self-reported ancestry. They found that blood levels of free circulating mtDNA were higher in the former group. Circulating mtDNA is potentially associated with cardiovascular diseases, autoimmune diseases, cancer, pregnancy complications, and neurodegenerative diseases.

Li et al.'s (2019) results suggest that mitochondrial genetic variation, particularly for the *MT-CO2* and *MT-ND2* genes, might play a role in breast cancer risk among Americans of European descent, but not in other geographical populations.

Reid et al.'s (2022) study indicates that Mexican American women have an increased risk of mtDNA damage, damage that is likely linked to comorbidities such as cardiovascular disease and diabetes. mtDNA damage could contribute to an increased risk of Alzheimer's disease in this ethnic group. Reardon (2023) noted that Hispanics and African Americans have a risk of Alzheimer's disease twice as high as Americans of European descent. In Reid et al.'s study (2022), it is difficult to know whether the higher incidence of mtDNA damage is mainly due to comorbidities (partly influenced by environmental factors) or the intrinsic properties of the mtDNA.

The incidence of ovarian cancer is higher among African American women than among American women of European descent. The five-year survival rate is 38% for the former versus 45% for the latter (Shukla and Singh, 2021). The authors note that comorbidities and living conditions likely play a significant role in these differences. However, they also suggest that aberrations of mtDNA and mitochondrial proteins encoded by nuclear DNA likely contribute to these incidence differences.

Choudhury and Singh (2017) think that mtDNA plays a role in the different disease incidences among geographical populations. They note that the frequency of cancer is higher in African Americans than in other ethnic populations. Besides socioeconomic, educational, cultural, and environmental factors, they hypothesize that mtDNA and mitochondrial proteins encoded by nuclear DNA might also play a role in these differences.

Mendelian or Monogenetic Diseases

This term comes from the name of the Czech monk Gregor Mendel (1868–1884), a pioneer of genetics with his patient study of smooth and wrinkled peas. It designates genetic diseases caused by a single gene or a small number of genes. They are not the most frequent genetic diseases. In some cases, the disease is caused by a single mutation. This is the case with sickle cell anemia* (see below).

> *Mendelian diseases are caused by a single gene or a small number of genes.*

Mendelian diseases have the particularity of having particularly marked geographical distribution inequalities (Jorde and Wooding 2004). Single Nucleotide Polymorphisms (SNPs) specifically associated with Mendelian diseases tend to be low-frequency or rare variants (Fu et al., 2013). This implies that they must have very unequal geographical distributions, as is generally the case for low-frequency and rare variants, as we have already seen (see: "The contribution of low-frequency and rare genetic variants"). There is no satisfactory explanation for the unequal distribution of these Mendelian diseases among geographical groups.

Sickle Cell Anemia

Sickle cell anemia (drepanocytosis) is very illustrative for the subject of this book. The disease results from a single mutation on one of the genes controlling hemoglobin* synthesis. The maintenance of this genetic variant in current natural populations is explained by a case of "balanced selection*": homozygous* carriers are affected by severe anemia, while heterozygotes* are not ill and show some resistance to malaria. This explains why this genetic trait persists in regions heavily infested by the parasite that causes this disease, primarily intertropical Africa, where, in some countries, the transmission rate is very high. Sickle cell disease is also observed in North and South American subjects of African descent. However, in the United States, among populations of African descent, its frequency is slightly lower than that observed in intertropical Africa (Gentillini et al., 1986). This might stem from the fact that African Americans have been protected from the selective pressure of malaria for several centuries. This shows that evolution continues to act in the present era, and on brief timescales. Sickle cell anemia is a classic case of a Mendelian disease with unequal geographical and ethnic distribution. It is not limited to intertropical Africa. It is also observed in certain regions of India, the Middle East, Saudi Arabia, and, exceptionally, Turkey, Greece, and the Maghreb (Gentillini et al., 1986). But its frequency in Africa is particularly high. The genetic variant associated with sickle cell disease has a frequency 1500 times higher in Africans than in Europeans (The UK Biobank Whole-Genome Sequencing Consortium, 2025).

Hereditary Hemochromatosis and Cystic Fibrosis**

These two Mendelian diseases are both much more frequent in populations of European descent than in other geographical groups (Armengol et al., 2009; González Burchard et al., 2003; Jorde and Wooding, 2004; Maglo, Mersha and Martin, 2016; Mersha and Abebe, 2015; Mountain and Risch, 2004; Risch et al., 2002; Wang et al., 2024). The first of these two diseases is caused by the storage of iron in different organs, leading to tissue damage. Cystic fibrosis is caused by mutations in the *CFTR* gene that lead to the accumulation of mucus in the pulmonary alveoli, resulting in respiratory failure. According to Gurdasani et al. (2019), the mutations involved are not the same in patients from different geographic populations. The P.Phe508del mutation is found in 90% of patients of European ancestry, but only in 17%, 30%, 38%, and 40% of Native American, Latino, African, and Asian patients, respectively.

Jordan (2008) hypothesized that cystic fibrosis might have been maintained by a phenomenon of balanced selection, like sickle cell disease, due to this genetic trait having provided some protection against cholera.

*Phenylketonuria**

This disease is caused by a mutation that leads to major metabolic disorders and mental retardation in affected children. It can be partially controlled by a suitable diet. According to Hardelid et al. (2008), in the United Kingdom, the disease is most frequent in children of European descent, then in children of Asian origin, and finally in children of African origin.

Huntington's Disease

Huntington's disease is a Mendelian neurodegenerative condition seen almost exclusively in Europeans (Winther, 2018).

Neurodegenerative Diseases Frequent in Ashkenazi Jews

Ashkenazi Jews (ancient Jewish population of Russia, Eastern Europe, Germany, eastern France) represent a fascinating case concerning certain neurological diseases. This population has the particularity of having an IQ* 15 points higher than that of other Europeans, according to Cochran and Harpending (2009). These authors speculated that Ashkenazi Jews paid a high genetic cost to benefit from these specific intellectual abilities. Through the interplay of balanced selection, this high IQ of Ashkenazi Jews would be associated with a heavy burden of Mendelian neurological diseases (Tay–Sachs, Gaucher, Niemann–Pick, mucolipidosis type IV). Ashkenazi Jews specialized in finance since the Middle Ages, as they were forbidden from owning land in those ancient times. These money-related professions, in a world without computers and artificial intelligence, demanded high intellectual capabilities. This fact, associated with endogamy* (the tendency to marry within one's own group), stimulated by persecution as well as tradition, and an original founder effect* (the Ashkenazi community having been founded by a relatively restricted number of individuals), is thought to have caused strong selection favoring the genes believed to underlie high IQs. This selection process, according to Cochran and Harpending (2009), is the cause of the high percentage of neurological disorders in this community. This hypothesis has been strongly criticized by Lala and Feldman (2024), who see it merely as "just-so-stories," but without presenting evidence to refute the hypothesis.

Regardless of the value of this audacious yet debatable hypothesis, the frequency of these neurological disorders in Ashkenazi Jews is a medical classic, like the frequency of congenital hip dislocation in Bretons. Ashkenazi Jews also exhibit high rates of cystic fibrosis (Guha et al., 2012).

The Ashkenazi Jewish community, which is genetically quite homogeneous, has been at the forefront in the field of personalized medicine, promoting effective screening for Mendelian diseases and breast cancer (Ostrer and Skorecki, 2013).

As in the case of the Fulani people and susceptibility to malaria (Arama et al., 2015), the example of Mendelian diseases in Ashkenazi Jews shows that in many cases, the main geographical groups (Europeans, Africans, Asians) are too broad units of analysis for the study of diseases (Barbujani and Colonna, 2010).

"Complex" Diseases

Complex diseases, also called "common diseases," are more frequent than Mendelian conditions.

> *Predisposition to complex diseases, or common diseases, is caused by the action of many genes, each having a limited individual effect. The environment plays a major role in these diseases, interacting with the genetic background.*

These diseases result from the action of many genes, each having a limited individual effect. Furthermore, the environment (lifestyle, diet, socioeconomic factors, education, access to care) most often plays an important role in the onset of these diseases. Modern genomics makes it possible to distinguish the genetic and environmental factors underlying complex diseases. Cardiovascular diseases and type 2 diabetes, for which lifestyle is a key factor, are typical examples of these diseases. Psychiatric and autoimmune disorders can also be cited, these two pathological categories seeming to share some genetic background (see further: "Psychological disorders, mental illness, and geographical populations").

Like Mendelian diseases, complex diseases are unevenly distributed across the world, with a variable impact depending on the major geographical groups (Tishkoff and Kidd, 2004). This is likely due to the fact that their heritability primarily relies on rare or low-frequency variants, while common variants would play a more limited role (Tennessen et al., 2012). As already discussed, these rare or low-frequency variants also tend to be specific to certain geographical regions. Here are a few examples illustrating the unequal distribution of complex diseases.

High Blood Pressure (Hypertension)

African Americans are more affected by this condition than Americans of European descent. It is evident that lifestyle influences this, as it differs notably between these two communities. However, even accounting for the environmental impact, genetic factors remain important (Tibayrenc, 2017b). Large differences in the frequencies of blood pressure-related genes have been observed between Africans and Europeans (Gurdasani et al., 2015).

> *The "slavery hypothesis" has been proposed to explain the preponderance of high blood pressure in African Americans. It remains speculative.*

The "slavery hypothesis" has been suggested to account for the higher prevalence of high blood pressure in African Americans (Brothers et al., 2021). According to this hypothesis, individuals who retained sodium more easily were positively selected during the long and grueling journey from Africa, which would have allowed them to better cope with the adverse conditions of deportation, by reducing diarrhea and vomiting. But this increased sodium retention would be the source of the current prevalence of high blood pressure in this community. This hypothesis obviously remains speculative.

According to Graves (cited by Gaskin, 2025), genes that lead to high blood pressure are more common in Americans of European descent than in African Americans. According to this author, the higher prevalence of high blood pressure in the latter group is therefore due to the environment. However, according to Gurdasani et al. (2015), high blood pressure is also more prevalent in Africans from Africa, who live in a completely different environment. This suggests that the genetic component of this trait is notable in populations of African descent.

Dilated Cardiomyopathy (DCM)

African Americans are disproportionately affected by this condition. Huffman et al. (2025) attributed this to the presence of a genetic variant (rs3211938:G), which occurs in 17% of African Americans, compared to less than 0.1% of Americans of European descent. Individuals homozygous for this genetic variant have a threefold increased risk of dilated cardiomyopathy.

Type 2 Diabetes Mellitus (T2DM)

Type 2 diabetes is distinct from Type 1 diabetes (insulin-dependent), the latter being an autoimmune disease that primarily affects children. Type 2, on the other hand, is polygenic and strongly influenced by lifestyle, particularly dietary hygiene, physical exercise, and overweight.

Type 2 diabetes is strongly influenced by the environment but has a genetic component. Its geographical distribution is highly unequal.

As with high blood pressure, its distribution varies among human groups. The risk is highest in Africans, followed by Europeans, Middle Easterners, and finally Asians (Corona et al., 2013). The origin of these differences is debated. The "thrifty gene hypothesis" (Neel, 1962) offers a classic explanation.

The "thrifty gene hypothesis," which is highly controversial, proposes that our ancestors developed a strong ability to store lipids to cope with starving periods.

According to this theory, our hunter-gatherer ancestors, undergoing frequent periods of famine between two luckily and randomly killed mammoths, would have developed a strong ability to store lipids, a vital asset at the time, allowing survival in lean times, but problematic in the current context of food overabundance. This would favor obesity and associated pathologies, including type 2 diabetes. Cochran and Harpending (2009) criticized this theory, believing that our hunter-gatherer ancestors did not experience famine. It should be noted, moreover, that present-day hunter-gatherers are not undernourished. However, they do not face the climatic rigors endured by our ancestors in Europe during the glacial periods. All these explanations undoubtedly remain highly speculative.

Cochran and Harpending (2009) highlight that the rate of type 2 diabetes is four times higher in Australian Aborigines and 2.5 times higher in Navajo Indians than in Europeans. This difference could be explained by an incomplete adaptation of these populations to an agricultural diet, which is richer in carbohydrates and poorer in protein than that of hunter-gatherers.

González Burchard et al. (2003) observed that among Pima Indians of Arizona, who have a very high prevalence of type 2 diabetes (38%; Gurdasani et al., 2019), a greater proportion of European genetic contribution, assessed by self-reported ancestry, is correlated with better protection against type 2 diabetes.

Whatever the precise genetic causes underlying this fact, the frequency of type 2 diabetes clearly varies from one geographical population to another.

Asthma and Respiratory Functions

Respiratory diseases clearly illustrate the importance of considering patients' geographical origin during medical examinations. Physicians use a device called a "spirometer" to assess patients› lung function. Contrary to the assertion of Omiye et al. (2023), lung capacity differs between geographical populations (see further: "Recent evolution in humans/adaptation to altitude"). Values considered normal for a given geographical group will be deemed pathological and signs of a lung disease for another group (Moreno-Estrada et al., 2014). Mersha and Abebe (2015) noted that the level of European genetic contribution in Mexican American individuals is the cause of greater severity of asthmatic affection. According to Vasseur and Quintana-Murci (2013), ethnic origin has a greater impact than the environment on the frequency of asthma. These authors suggested that populations from tropical regions living in northern countries are more affected by asthma. This would be the case for African Americans.

Inflammatory Bowel Diseases (IBD)

This term actually encompasses two conditions: Crohn's disease and ulcerative colitis. These diseases result from a dysregulation of the immune system. The contributing factors are genetic, epigenetic, and microbiomic (Hao et al., 2025). IBDs have a different prevalence in different parts of the world: in descending order of prevalence, first Oceania and Europe, then North America, then Asia, the Middle East, and South America (Hao et al., 2025). Environment and lifestyle certainly play a significant role in their development, as suggested by the fact that they are more common in urban areas. Several genetic factors show differences in frequency between geographic populations: The SNPs *ZNF649* and *LSAMP*, associated with IBD, are specific to populations of African descent. The *NOD2* risk gene has a frequency of 13% in Europeans compared to 0.05% in Southeast Asia. The *ATG16L1* risk variant is more frequent in Europe than in the rest of the world. The *TNFSF15* variant increases the risk of IBD more in Southeast Asians than in Europeans. Generally speaking, there is a difference between Europe and the rest of the world regarding genetic risk factors for IBD (Hao et al., 2025).

Cancer

Cancers show a difference in distribution between geographical populations, differences that are partly of genetic origin.

Cancer, or rather cancers, is (are) a specific case of complex disease, in which the environment plays a large role, more or less important depending on the type considered, and in which differences in distribution between geographical groups are noted.

In a very exhaustive review, Thomas and Peters (2025) underscore the close entanglement of genetics, environment, and social conditions in the differences in cancer distribution between geographical populations. They insist that this research is hampered by the constant overrepresentation of samples of European descent, a veritable leitmotif in this field of research. Despite this obstacle, these authors reach the very clear conclusion that the differential geographical distribution of many cancers has partly genetic causes.

Melanoma

According to Harris (2015), the frequency of cutaneous melanoma (an extremely malignant tumor) in admixed "Latinos" is proportional to their share of European ancestry. The frequency of melanoma is much lower in Africans compared to Europeans, suggesting climatic adaptation (defense against intense sun exposure) in Africans (Gentillini et al., 1986). It is obvious that the environment also plays an important role: the craze for excessive tanning is a well-known risk factor, with its accompanying repeated sunburns in youth. However, all else being equal, Europeans, especially those of Nordic descent, blond with blue eyes, pay a heavier toll for the same degree of UV exposure than other geographical populations.

Breast Cancer

Jia et al. (2025) find a difference in the distribution of genes predisposing to breast cancer between European and Asian women.

According to Martini and Davis (2024), African American women have 40% more breast cancer than American women of European descent, given equal socioeconomic conditions.

Ostrer and Skorecki (2013) note that three mutations in the *BRCA* 1 and 2 genes exhibit a high frequency in Ashkenazi Jewish women and confer a predisposition to breast and ovarian cancers. The presence of this genetic trait opens the way for early screening with possible preventive ablation, an illustrative case of personalized medicine, a field in which Israeli medicine is a pioneer. The same fact is noted by Thomas and Peters (2025). The latter authors highlight an interesting fact: the presence of the "minor" allele (i.e., the one with the lowest frequency) of a certain genetic variant (6q25 {rs140068132}) associated with an increased risk of breast cancer, leads to a decreased risk in Hispanic women. This classification category has frequently been criticized for its heterogeneous and artificial nature. Nevertheless, statistically, from a genetic point of view, the Hispanic population as a whole is different from African Americans, Americans of European descent, Asians, and Native Americans.

According to Vasseur and Quintana Murci (2013), the *UGT2BA* gene causes a higher risk of breast cancer in Nigerian women, and to some extent, in African American women.

Prostate Cancer

Verma et al. (2024) note a greater incidence of prostate cancer in Africans (identified by self-reported ancestry).

In a multi-ethnic study on the incidence of prostate cancer whose sampling included Africans, Europeans, and Asians, Jaratlerdsiri et al. (2022) note that traits specific to African ancestry include a high tumor mutational burden, an increased percentage of genome alteration, and a greater number of potentially damaging mutations. In their study, prostate cancer also had a higher frequency in Africans than in the other geographical populations.

Hoffmann et al. (2025) conducted a genome-wide association study on a vast sample including 211,342 Europeans, 58,236 Africans, 23,546 Hispanics, and 3,630 Asians. The polygenic risk scores for prostate-specific antigen (PSA), used as a screening test for prostate cancer, were different between the four populations examined.

The *HOXB13* gene, involved in prostate cancer risk, exhibits different mutations in Scandinavians, Chinese, Japanese, and African Americans (Thomas and Peters, 2025).

Colon Cancer

Thomas and Peters (2025) note that Alaskan Native populations have a significantly higher rate of colon cancer than other populations, including Native Americans, who are closely related to these Alaskan Natives. This underscores the need to consider smaller populations than the main geographical groups in many cases of genetic studies.

A variant of the *APC* gene, involved in an increased risk of colon cancer, is found in 6% of Ashkenazi Jews, while it is virtually absent in other populations (Thomas and Peters, 2025).

Wang et al. (2024) note that the incidence of colon cancer is 20% higher and its mortality is 40% higher in African Americans compared to Americans of European descent. These authors postulate a strong hereditary component for colon, ovarian, prostate, pancreatic, and endometrial cancers.

Lung Cancer

Betti et al. (2025) note that African Americans are affected by lung cancer "disproportionately," even though their cigarette consumption is lower than that of Americans of European descent. These authors identified several susceptibility loci for this disease that appear specific to African Americans.

Different Cancer Types

Amuzu et al. (2025) identified numerous associations of mutations linked to 14 different types of cancer, with notable variations in frequency across different geographic groups. The TP53 mutation, associated with many cancers, has an increased frequency in Africans and East Asians. The same mutation has a decreased frequency in these same geographic populations in the case of prostate adenocarcinoma. There is an increased frequency of the ERBB2 mutation, associated with lung cancer, in mixed-race American patients, independent of tobacco use. The CDK12 mutation, linked to prostate cancer, is more frequent in East Asian populations. Finally, in Africans, there is an increased frequency of the BAP1 mutation (head and neck cancer) and a decreased frequency of the FGFR2 mutation (uterine carcinoma). According to the authors, these results suggest geographic population-specific biological mechanisms in the development of cancers.

Psychological Disorders, Mental Illness and Geographic Populations

This section concerns only mental traits that are generally considered pathological, regardless of the difficulty in grasping the exact meaning of this concept. Indeed, the boundary between "normal" and "pathological" is sometimes blurred with respect to brain function, as masterfully demonstrated by Frances (2013).

The boundary between normal and pathological is difficult to define for mental disorders.

In fact, the thresholds defining psychiatric illness, as established by the Diagnostic and Statistical Manual of Mental Disorders (DSM-5), are arbitrary (Hyman, 2014). This is the case for many, if not all, thresholds used in medicine, such as the cutoff values for high blood pressure. However, high blood pressure is a much simpler parameter to define than schizophrenia or autism, and its epidemiological risks have been meticulously studied.

The heritability of major psychological disorders is far from negligible: it is 35% for depression, up to 65–80% for schizophrenia (Hyman, 2014), and 80% for autism, which is equal to that of height (Pearson, 2025). The search for genetic predisposition variants for these disorders is difficult due to the imprecise definition of many syndromes. However, this research offers hope for finding new treatments (Hyman, 2014).

Genetic predispositions to certain major mental disorders have been highlighted by genome-wide association studies.

Associations between given genetic variants and the increased risk of schizophrenia, autism spectrum disorders (ASD), major depression, and bipolar disorder (BPD) have been fully confirmed by genome-wide association studies (GWAS). In most cases, the loci identified by these GWAS concern regulatory regions rather than coding genes (Geschwind and Flint, 2015). It is likely that many genetic predisposition variants are shared between several mental disorders, as is the case between autoimmune and inflammatory diseases (Hyman, 2014). It appears that genes for inflammation and immune defenses are associated with those for psychiatric illnesses, including schizophrenia and autism (Karlsson et al., 2014). Genetic discoveries blur the boundaries between psychiatric illnesses, but also the limits between normal and pathological (see above). In fact, it seems that the polygenic risk associated with bipolar disorder and schizophrenia is also linked to creativity. In patients' families, there are individuals who are not ill and who exhibit intermediate phenotypes ("endophenotypes") (Geschwind and Flint, 2015).

The identification of the precise genetic background of mental illnesses could lead to their nosological redefinition, moving from a symptomatic approach to a causal approach (Cross-Disorder Group of the Psychiatric Genomics Consortium, 2013; Geschwind and Flint, 2015).

The present book focuses on the geographical diversity of our species. Unfortunately, the search for variations in specific psychological disorders among geographic populations has not been very fruitful. As is very generally the case in genomic research, samples lack diversity. "The vast preponderance of psychiatric genetic data is from Europeans, leveraging Scandinavian population registries and European national health systems" (Steven Hyman, personal communication). Empirical presumptions of differences in the prevalence of mental illnesses between geographic populations are available (Harris et al., 2005; Williams et al., 2007). The difference in the prevalence of mental illnesses between groups of ancestry is confirmed by data from the U.S. Centers for Disease Control and Prevention (CDC): see http://www.cdc.gov/mentalhealth/basics/burden.htm. However, these raw data do not constitute proof of genetic causes for these prevalence differences, as evident environmental parameters (sociological, economic, family, educational) certainly play a major role in the etiology of mental illnesses.

Few reliable data are available regarding differences in genetic predisposition to mental disorders between geographic populations, except for schizophrenia and Alzheimer's disease.

In the following two cases, however, among others, genetic data are available.

The genetic background of schizophrenia appears complex. Rare deletions (Lander, 2011), common variants (Lander, 2011; Zietsch et al., 2015), and microdeletions at specific loci on chromosomes 22q11.2, 1q21.1, and 15q13.3 have been implicated (Belizário, 2013), as well as rare CNVs (Gonzaga-Jauregui et al., 2012). No common CNV is known to be associated with schizophrenia (Zietsch et al., 2015). The Dystrobrevin Binding Protein 1 (*DTNBP*1) gene on the 6th chromosome has been shown to impact schizophrenia risk and also to influence general cognitive ability (Burdick et al., 2006). Regulatory phenomena are also involved in the etiology of schizophrenia (Geschwind et al., 2012). Finally, genes of the inflammatory and immune systems are associated with schizophrenia (Karlsson et al., 2014). Regarding differences between geographic populations, large genome-wide association studies (GWAS) relying on the common variant tagging suggest that many schizophrenia risk alleles are shared across ethnic groups and predate the Africa-Europe divergence (de Candia et al., 2013). However, this study only covers common variants and does not consider the possible role played by rare and low-frequency variants, which tend to have very different frequencies from one geographic population to another. Common variation is considered to account for more than 33% of the heritability of schizophrenia (Lander, 2011).

Alzheimer's disease is known to exhibit unequal frequencies among populations of different ancestries (Armengol et al., 2009; Farrer et al., 1997). The Apolipoprotein E (*APOE*) gene and its variants play a major role in the disease's etiology. The *APOE*ϵ4 variant is present in all geographic populations, albeit with different frequencies: Japanese: 9%; Europeans: 14%; African Americans: 19% (González Burchard et al., 2003).

Homozygosity for the ϵ4 allele increases the risk of Alzheimer's disease by a factor of 33 in Japanese individuals, 15 in Europeans, and 6 in African Americans. For heterozygous subjects, the Figures are 5.6, 3.0, and 1.1, respectively (González Burchard et al., 2003). Rare and common CNVs have also been implicated in the etiology of Alzheimer's disease (Armengol et al., 2009; Gonzaga-Jauregui et al., 2012). If rare CNVs play a role in Alzheimer's disease, this should lead to an increased differential geographic distribution of the disease, as rare variants tend to be population-specific.

Transmissible Diseases: A History of Co-evolution with Two or Three Actors

Generalities

Transmissible diseases are another specific case of complex diseases. They are caused by pathogenic agents. These pathogens can be: (i) viruses (e.g., dengue, flu, yellow fever, AIDS, SARS-coronavirus-19); (ii) bacteria (bacterial meningitis, tuberculosis); (iii) parasitic protozoa (malaria, toxoplasmosis, African human trypanosomiasis or sleeping sickness; American trypanosomiasis or Chagas disease); (iv) parasitic worms or helminths (schistosomiasis, onchocerciasis, tapeworm); (v) microscopic yeasts and fungi (candidiasis, cryptococcosis, aspergillosis).

Transmissible diseases offer an ideal model for evolutionists, and in return, evolutionary sciences have significantly contributed to the knowledge and control of these diseases. This was the theme of the scientific journal of which I was editor-in-chief from 2000 to 2021, Infection, *Genetics and Evolution* (Elsevier; http://www.elsevier.com/locate/meegid). Transmissible diseases are a case of co-evolution between the host and the pathogenic agent. In the case of vector-borne diseases, i.e., where the pathogenic agent is inoculated by a vector, such as malaria or African human trypanosomiasis, the co-evolution phenomenon involves an additional actor. Vectors are most often insects (e.g., the *Anopheles* mosquitoes in the case of malaria), but they can also be other arthropods (ticks, crustaceans) and mollusks.

> *In transmissible diseases, the host, the pathogen, and the vector (in the case of vector-borne diseases) are constantly adapting to each other: this is called co-evolution.*

This is called co-evolution because the host (e.g., humans), the pathogen, and the vector have had to adapt to each other over time. If the pathogen kills its host or makes it sick, it is not out of sadism; it is to survive. And in fact, its best-understood interest, to speak in teleological terms, is for the host to stay alive. As for the vector, it does not transmit the pathogen to do it a favor. A mosquito that bites you does not aim to "give" you yellow fever. It is simply a female taking a little blood for her oogenesis. Finally, the host has developed a whole sophisticated panoply of immunological defenses to try to survive the infection.

According to the so-called "Red Queen" hypothesis (Van Valen, 1973), the name of which is inspired by Lewis Carroll›s novel "Through the Looking-Glass", our species is engaged in a constant race against pathogens, and, of course, against vectors. "It takes all the running you can do, to keep in the same place."

The Considerable Role of the Environment

> *The environment plays a considerable role in the transmission and severity of infectious diseases, especially socioeconomic factors ("diseases of poverty").*

As is the case for other complex diseases, the development of transmissible conditions is strongly conditioned by the environment. This includes education, customs, nutrition, access to healthcare systems, economic factors, political instability, and wars.

"Diseases of Poverty"

Malaria, AIDS, and tuberculosis are called "diseases of poverty" because these conditions particularly affect low-income countries. Similarly, Chagas disease, which is observed from the Southern United States to Northern Argentina, is an illustrative example of the influence of socioeconomic factors on the impact of a disease. I have been studying this disease for 45 years (Figure 1 in the introduction). The causative agent is *Trypanosoma* cruzi, a parasitic protozoan related to the agent of African human trypanosomiasis. This parasite is transmitted by blood-sucking bugs (haematophagous*) of the Triatominae subfamily (Figure 10), called vinchucas (a Quechua word) in the Andean countries. The unappetizing North American name is "kissing bug," because these impressive insects, which can be several centimeters in size, have the appealing habit of biting at the corner of the mouth, where the skin is softer.

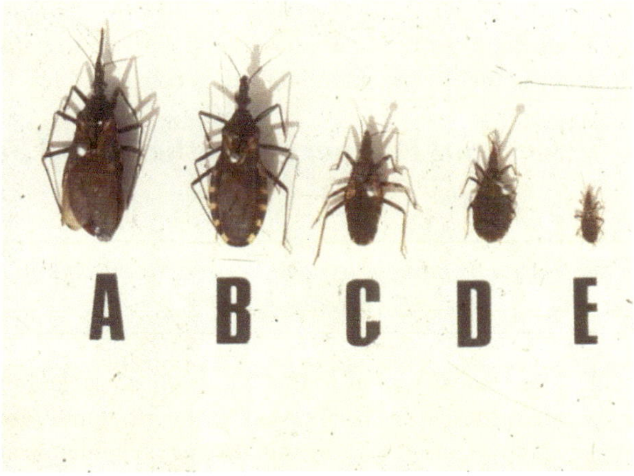

Figure 10: Vinchucas, here of the species *Triatoma infestans*, very widespread in human dwellings in South America. From left to right: male, female, and larvae. Both adults and larvae are strict blood-feeders. Note the impressive proboscis ("trunk") of the male, which can pierce any mosquito net. The icing on the cake is that the bite is painless: no warning signal.

These unfriendly insects proliferate in poor habitats in Latin America. They favor traditional constructions made of "adobe," a material made of clay simply dried in the sun (Figure 11).

These adobe houses are the lot of poor populations in South America. This is why Chagas disease has been called a "disease of underdevelopment." It is no coincidence that transmission in homes in the United States is exceptional: North

Figure 11: A traditional habitat made of "adobe" in Bolivia. The walls are full of holes and cracks in which the "vinchucas" hide, coming out at night to take their due from the sleepers.

American houses have nothing to do with the one shown in Figure 11. In these "diseases of underdevelopment," or "diseases of poverty," the influence of the environment, particularly the socioeconomic level, is considerable.

Hunter-Gatherers and Farmers; Northern and Southern Countries

> *The advent of agriculture is thought to have favored the spread of transmissible diseases.*

A classic and quite logical notion is that the birth of agriculture favored the spread of transmissible diseases. In the time of Neanderthal and Cro-Magnon, our ancestors lived in small communities. If, by misfortune, an epidemic struck such tiny groups, it remained limited to those groups. This is what is observed nowadays among hunter-gatherers. When humans began to establish crops and raise pigs, chickens, and cows, they also built villages that housed communities based on the principle of task sharing. Promiscuity, less rich food, often deficient in protein, and intimate contact with domestic animals favored the spread of infectious diseases, particularly those with an animal reservoir ("zoonoses") (Cochran and Harpending, 2009). Of course, globalization and modern transportation (airplanes) worsen the situation, as shown by the rapid global spread of the COVID-19 epidemic. Haldane (1949) was the first to conceive that transmissible diseases have been the predominant factor in human selection over the last 5,000 years.

> *The selection pressure of transmissible diseases, for example, malaria, is still operating today.*

It would be a grave mistake to think that modern life protects us from this selection pressure from infectious diseases. Nowadays, this would be partially true in industrial countries, with the caveat that even in these areas, nosocomial infections (contracted in the hospital) kill hundreds of thousands of people each year. But these nosocomial diseases primarily strike elderly subjects who have passed reproductive age, and thus, this less concerns the phenomenon of natural selection. Conversely, in Southern countries, in Africa, for example, malaria, for instance, kills many very young subjects. There is no doubt that natural selection by infectious diseases is still at work in these regions, since survivors can transmit their genes, but not those who succumbed to the disease.

Geography Dictates Inequalities

People living in tropical Africa do not face the same infectious dangers as Europeans. These two geographic areas have radically different populations of pathogens and vectors.

> *Populations of pathogens and vectors vary considerably depending on geography and climate.*

Tsetse flies or *Glossina*, the vectors of sleeping sickness, currently only live in tropical Africa. *Anopheles* mosquitoes, which transmit malaria, are found in Europe as well as in Africa. However, the parasite that causes the most malignant form of malaria, *Plasmodium falciparum*, does not (yet?) circulate in Europe. Many other examples can be cited. Africans and Europeans have therefore had to adapt to different infectious challenges, and the same goes for other geographic populations.

Historical Factors

> *Military conquests, colonization, and slavery have played a major role in the spread of transmissible diseases. Native Americans were decimated by germs brought by the conquistadors, germs against which they had no immune defenses.*

History plays an important role in the circulation of infectious diseases, a direct consequence of the specific adaptation of different geographic groups to specific pathogens. During the conquest of Latin America by the Iberian conquistadors, the latter were numerically very inferior to the Native Americans. Furthermore, the local inhabitants were anything but defenseless. The Aztec and Inca armies were experienced in the art of war and ruled over vast empires. Of course, the conquistadors were equipped with muskets and blunderbusses. However, their ultimate weapon consisted of infectious germs from Europe. The Native Americans were immunologically defenseless ("naive") against them. Tuberculosis, smallpox, and influenza caused massacres in the invaded populations, far more than the primitive firearms of the conquistadors. Conversely, in other circumstances, it was the conquerors who were halted by local germs. It has been noted that the limit of the penetration of Islam in Africa corresponds more or less with the regions where the tsetse fly and sleeping sickness circulate. This observation led to the hypothesis that the Arab conquerors were stopped by trypanosomes, the parasites that cause sleeping sickness. More specifically, it may be that their horses, an essential element of the Arab armies, were decimated by the animal form of sleeping sickness.

French Guiana earned the mournful nickname of "white man's grave." Until modern life completely changed the situation, all attempts to populate this country with French settlers failed, while Native Americans and African slaves, better adapted to tropical diseases, resisted much better. Until 1946, French Guiana housed the infamous prison camp described very realistically in the film Papillon (1973). In addition to their cruel living conditions, the convicts paid a heavy price to tropical diseases, especially yellow fever.

Genetic Inequality Against Transmissible Conditions

The distribution and importance of pathogen and vector populations differ considerably between geographic areas. This has caused different selection pressures on the human populations concerned.

> *Our immune defenses against infectious diseases are unevenly distributed, both at the individual and population levels.*

We are genetically unequal in the face of infectious diseases, probably more so than for all other diseases. This is true at the individual level and at the population level. For the same infectious contact, all other environmental conditions being equal, one individual will contract tuberculosis, another will not. And populations of different geographic origins, in the same environment, will not react identically to the same disease.

How Do Humans Genetically Protect Themselves Against Pathogens?

Three major diseases illustrate human genetic adaptations against their tiny enemies:

> *Geographic populations are unequally vulnerable to malaria, sleeping sickness, and AIDS.*

Malaria and sickle cell anemia (drepanocytosis) have already been discussed (see: "Mendelian diseases"). Let us recall that subjects homozygous for the sickle cell trait present a severe disease, while heterozygotes are healthy and are protected from malaria. This is an illustrative case of balancing selection. Such protection is, of course, valuable in tropical Africa. Sickle cell anemia is still present among African Americans, even though this population is no longer exposed to malaria.

Sleeping sickness or African human trypanosomiasis is a calamity in intertropical Africa. Africans, unlike Asians and Europeans, have developed partially effective genetic tools to better resist this disease. However, African Americans, who are no longer threatened by this disease, are afflicted with a higher rate of kidney ailments than Americans of European descent. This has been attributed to variants of the Apolipoprotein L1 (*APOL1*) gene, which would provide some protection against African human trypanosomiasis (Karlsson, Kwiatkowski, and Sabeti, 2014). This would be another case of balancing selection, as with sickle cell disease.

AIDS, caused by Human Immunodeficiency Viruses 1 and 2 (HIV 1 and 2), is a rather counter-intuitive example of genetic protection against a disease, since Africans seem less resistant to this disease than Europeans. The explanation for this unusual fact would lie in a gene deletion called CCR5Δ32. This genetic trait is much more frequent in Europeans than in Africans (Dean et al., 1996; Gurdasani et al., 2019). Subjects heterozygous for this deletion have a slowed progression of the disease, and homozygotes are very resistant to infection by the AIDS virus (Chapman and Hill, 2012). According to McLaren et al. (2023), the *CHD1L* gene is involved in the HIV viral load specific to African populations.

Other Transmissible Diseases Unevenly Distributed by Geographic Group

> *Geographic populations exhibit different degrees of innate susceptibility to many other transmissible diseases, particularly SARS-COV2 infection.*

Cytokines* are biochemical messengers involved in immunological defense against infectious agents and in inflammatory processes. Boahen et al. (2022) highlighted numerous genetic differences related to cytokine metabolic pathways between a Tanzanian population and a Western European population. This is undoubtedly due to the fact that Africans and Europeans have long faced radically different pathogen populations.

Bucheton et al. (2002) found strong ethnic differences among Sudanese patients regarding the ratio of asymptomatic/symptomatic patients concerning visceral leishmaniasis, a parasitic disease with a severe prognosis.

According to Koch (2022), there are differences in genetic susceptibility for the influenza A virus and SARS-COV2 (COVID) between ancestry groups.

Differences in innate susceptibility between geographic populations to COVID infection have been implicated by some authors (Niedzwiedz et al., 2020; Shelton et al., 2021).

According to Quintana-Murci (2021), some of these differences related to COVID infection could be attributable to adaptive Neanderthal archaic introgression, the genes in question being present in 16% of Europeans, 50% of Indians, and 63% of Bengalis.

It is, moreover, quite evident that environmental and socioeconomic conditions as well as racial discrimination also play an important role in the affliction by SARS-COV2 (Graves and Goodman, 2022; Parens, 2020). As I have already emphasized, this must not lead to artificially opposing genetics and environment, as Graves and Goodman (2022), Parens (2020), Roberts (2021), and Yudell et al. (2020) do.

Conclusion on the Genetic Susceptibility of Humans to Diseases

The presentation given in this chapter is deliberately simplified and abbreviated. However, it is based on the most recent research findings and relies on the considerable technological advances that have enriched this field of research. Genome sequencing is now routine, and its cost has been lowered 10 million times in 20 years (Alexandre, 2024). All the most recent results point in the same direction: the genetic diversity of human populations is finely stratified, and this extends to very small geographical scales.

> *Past migrations have mixed humankind's genes without homogenizing them. Geographical populations remain stratified, down to a very fine scale, with geographic distance being the main causal factor in this pattern. Genetic markers can identify the geographical origin of any individual*

Abundant migrations have heavily mixed the genes of humankind but have not homogenized them. Humankind is by no means a "panmictic*" entity (within which genetic exchanges occur according to the laws of chance). The phylogenetic signal, although attenuated by genetic exchanges, persists, both at the intercontinental level and on a very reduced spatial scale. The local geographical scale resembles a miniature of the global image. The genetic structure of humankind thus exhibits a "Russian Doll pattern," a concept coined to describe the structure of pathogenic microorganism populations (Tibayrenc and Ayala, 2013). In the case of microbes, stratification is caused by biological restrictions on genetic exchanges: microbes "restrain their sexuality." Nothing of the sort exists in humans. There is rigorously no biological barrier to fertile reproduction between men and women of any population around the globe. In humans, stratification has been caused by geographical separation, as well as by linguistic obstacles and other cultural traits.

Genetic markers can identify the geographical origin of any individual with extreme acuity, and also their potential level of admixture. This very resilient pattern has extended applications in forensic medicine.

Furthermore, we have seen that human genetic diversity is of considerable importance in the prevention, epidemiological monitoring, and treatment of diseases. The future of this medical field must be based on more integrative approaches, taking into account both genetic and environmental factors simultaneously.

The rapid progress of modern technologies (proteomic analysis* of gene regulation and epigenetic phenomena, high-throughput sequencing, "cloud computing," or mega-informatics in general) paves the way for personalized medicine, finely targeted at specific individuals or populations. To achieve this, fields of knowledge that have been separated until now must be decompartmentalized and opened to each other: evolutionary sciences and biomedical research, biology and human sciences. Our knowledge of human diversity, whether for fundamental science or for medicine, must not be restricted to biology. This is all the more true as the environment in the broad sense (ecology, socioeconomic conditions, education) has a fundamental impact on human diversity, alongside biology.

As already emphasized (see above: "erroneous dichotomy: nature/nurture"), it is wrong, and even sophistical, to consider that taking into account the heritable part of diseases automatically leads (Duncan et al., 2024; Graves and Goodman., 2022) to neglecting the impact of the environment, known as the social determinants of health (Wang et al., 2024). Both must be taken into account in the genesis of diseases in an indispensable integrated approach. This is what classic medicine includes in the concept of health status, which encompasses not only genetics but also age, lifestyle, social setting, and the environment. I have coined the term "socioecology" to summarize all these complex non-genetic factors. This integrated approach is the specific theme of the new scientific journal I am the editor-in-chief of: *Disease biology, genetics and socioecology* (https://www.sciltp.com/journals/dbgs/).

Some authors (Brown et al., 2025) exclude any innate component in the differences in pathology distribution between African Americans and Americans of European descent and attribute these differences entirely to the stress due to systemic racism and the resulting epigenetic changes. This proposition is certainly to be considered, but the results presented above suggest that it is excessive.

> *Personalized medicine will have to rely on close collaboration between biology and social sciences.*

In the same vein, it has been claimed that evaluating differences in genetic susceptibility to diseases between geographical groups perpetuates "structural racism" (Westby et al., 2020). This evaluative approach that takes into account genetics has been caricatured by saying that it amounts to considering sickle cell disease as a "Black disease" and cystic fibrosis as a "White disease" (Yudell et al., 2016). No one has ever claimed, as Deyrup and Graves (2022) imply, that a given disease-related genetic trait was present in 100% of a given geographical group and totally absent in other groups.

Chapter 3

Some Illustrative Case Studies

What Are Europeans Made Of?

Europeans, the original inhabitants ("natives") of the European continent, exhibit high phenotypic diversity: from the towering, blonde, blue-eyed Dutch (the tallest people in the world, according to the latest news) to the stocky, dark-haired Mediterraneans. However, they are immediately categorized as such ("racialized") by people from other continents. In Thailand, we are the "farangs," a non-discriminatory but strongly identifying term. In Bolivia, we are the "gringos."

> *The original inhabitants of Europe exhibit high phenotypic diversity but are statistically distinguished from other continental groups by genetic markers.*

From a genetic point of view, there is no ambiguity: European ancestry at the population level is reliably identified by modern genetic markers. This is clearly visible in graphical visualizations of "genetic distances" (Figures 4–8). This does not mean, as the misleading typological model would suggest, that the "European" geographical group is genetically homogeneous, nor that it is clearly separated from other geographical groups. Neither of these two propositions is verified. If one refers to "classic" genetic markers (those that were massively used from the 1970s onwards), the European group has the particularity of having the lowest genetic variability, much lower than that of Africans, for example.

Who Are the Ancestors of Europeans?

According to current data, one of the oldest known Europeans is *Homo antecessor*, found at the famous Spanish site of Gran Dolina (Great Sinkhole), whose age has been estimated at 772,000–949,000 years, i.e., well before the "out of Africa" event (Bordenave, 2018a).

DOI: 10.1201/9781003647393-4

But there is better: a fossil discovered at the same site has been dated to – 1.1 million years ago and has been attributed to *H. erectus*, or one of its close relatives (*Homo aff. erectus*) (Gibbons, 2025; Huguet, 2025). This is currently the oldest human remains discovered in Western Europe.

If we turn to Eastern Europe, the Dmanisi site in Georgia has yielded numerous well-preserved fossils dated to 1.8 million years ago. These fossils have been attributed either to *H. erectus* (formerly called "Pithecanthropus": the ape-man), to *H. georgicus* (Lordkipanidze, 2017), or even to *H. habilis* or *H. ergaster* (Bettayeb, 2025). These hesitations underline the difficulty of defining paleo-species. The Dmanisi fossils show the antiquity of the first migrations out of Africa.

Two famous, less ancient fossils are the Mauer man (Germany) and the Tautavel man (Pyrénées-Orientales, France), 400,000 to 600,000 years old, both attributed to the species *Homo heidelbergensis*. The settlement of our continent is therefore very old, far predating the appearance of *Homo sapiens* and even *H. neanderthalensis*.

The latter, the "Neanderthal man" (from the Neander Valley in Germany, where the first fossil of this species was discovered in 1856), populated our continent, as well as the Middle East and Central Asia, for a very long time, from about 400,000 years ago. It died out 30,000 years ago. Hazardous hypotheses suggest the survival of lingering Neanderthal populations today in remote corners of the planet (Heuvelmans and Porchnev, 1974). This sensationalist hypothesis, which delights conspiracy theorists, has not received reliable confirmation.

Neanderthal, therefore, coexisted with our *sapiens* ancestors, with whom, we are now certain, they interbred. Genetic exchanges between the two species, contrary to what was long claimed (Currat and Excoffier, 2004; Jordan, 2008), did indeed take place. It even appears that such genetic exchanges occurred repeatedly, at various times and places (Callaway, 2021). However, this hypothesis has recently been questioned (Iasi et al., 2024).

Modern non-African populations have approximately 5% Neanderthal genes

Europeans and Asians, but not Africans, thus harbor about 3–5% (depending on the estimates) of Neanderthal genetic heritage. And since this "archaic" heritage is not the same from one individual to another, some 30% of Neanderthal genes persist in current non-Africans, which contributes to the genetic differentiation between non-Africans and Africans. Furthermore, the Neanderthal contribution is thought to be less important in West Eurasians than in East Asians (Bon, 2025).

Rather than speaking of "disappearance", it would be more accurate to say that the Neanderthals merged into the *sapiens* immigrant. As we have seen (see "origin of human races"), this "adaptive archaic introgression" (Gaudieri and John, 2017; Gunz et al., 2019; Stoneking, 2017; Stoneking and Krause, 2011) has had many phenotypic consequences, on adaptation to cold climates, or on disease resistance. Neanderthal is therefore undoubtedly part of the genetic heritage of Europeans.

The existence of genetic exchanges between *H. sapiens* and *H. neanderthalensis* questions the taxonomic status of the latter. Species? Subspecies? (Dignat, 2018). In the latter case, it would receive the triname *H. sapiens neanderthalensis*. The question is not settled. However, the status of a distinct species generally prevails (Wood and Boyle, 2017).

Our archaic ancestor was not the ape-like brute described by authors of the *Belle Époque*, such as the famous French prehistorian Marcellin Boule. He had elaborated funerary rites, as evidenced by the burial of the man of La Chapelle-aux-Saints in France. He cared for the elderly and the disabled, developed sophisticated lithic industries, and, as is now thought, was also an artist: we have no reason to be ashamed of our robust ancestor.

It is generally accepted that *H. neanderthalensis* and *H. sapiens* separated about half a million years ago (Gibbons, 2017a; Hublin, 2011).

> *"Anatomically modern humans" (phenotypically indistinguishable from present-day humans) have been present in Europe for 45,000 years.*

H. sapiens is sometimes referred to as "anatomically modern human." This means that if you crossed him on the street with current clothes and haircut, you would not notice anything particular. So, *H. sapiens* is you and me.

The first discovery of a fossil of this species is the famous "Cro-Magnon man". Cro-Magnon is a locality in the commune of Les Eyzies de Taillac-Sireuil (Dordogne, France). The "Cro-Magnon man" was discovered in 1868. He would have lived about 28,000 years ago. He is therefore not the oldest *H. sapiens* in Europe. That honor would go to a specimen recently discovered in Bulgaria, which would have lived some 45,000 years ago (Hublin et al., 2020; Lewino, 2020).

These ancient European *H. sapiens* are probably in part the direct ancestors of Modern Europeans. Indeed, the analysis of ancient DNA has shown that current European populations have three main components: steppe pastoralists (Yamnaya people, corresponding to ancient North Eurasians), ancient European farmers, and Western Mesolithic* *hunter-gatherers* (Haak et al., 2015). The latter were thus visited by the Yamnaya and farmer immigrants, not necessarily peacefully. But ultimately, all these people mixed their genes, without population substitution. The proof: we are here, with Yamnaya, farmer, and hunter-gatherer flowing in our veins.

> *Modern Europeans have three population components: original hunter-gatherers, ancient farmers, and "steppe peoples" (Yamnaya).*

The contribution of each of these three groups is unevenly distributed among different modern European populations. The steppe/Yamnaya input is more limited in Southern Europe, while the hunter-gatherer component is more important in the North (Haak et al., 2015). This is what Europeans are made of.

Some ancient DNA data (Allentoft et al., 2024; Leroyer, 2024) suggest that these population substitutions occurred quite violently, at least south of Denmark, with the near elimination of hunter-gatherers by Anatolian farmers, and then of the latter by steppe pastoralists. However, in a major part of Europe, the three populations contributed to the genetic makeup of current Europeans, in varying proportions.

This admixture has repercussions on important phenotypic traits such as height, skin pigmentation, heart rate, platelet count, bone mineral density, and many other characteristics. These phenomena have been found in the study of two different genetic banks, the UK Biobank and the Estonian Biobank. It is therefore assumed that they are generalizable to all of Europe. The environment is not the cause of this data (Pankratov et al., 2024). This stratification due to our three ancestral components must therefore be taken into account for all studies of phenotypic traits, including those of medical interest. This is the triple genetic heritage of our ancestors.

Genetics of Modern Europeans

According to Harris (2015), it is possible to identify 462,876 "private" (i.e., specific) European SNPs. These markers turn out to be variable in European populations, while they are fixed (i.e., non-variable) in Africans and Asians. This denotes an undeniable genetic specificity of Europeans.

Other results corroborate the fact that Europe is far from having a genetically homogeneous population.

According to Jordan (2010), the use of numerous SNPs makes it possible to differentiate French, Belgians, and Swiss.

According to Novembre et al. (2008), the genetic structure of Europe does not show discontinuities but is not homogeneous. It evokes a continuum, reflecting geographical distance. Cultural peculiarities also seem to play a role: within a relatively small country like Switzerland, genetics makes it possible to differentiate German-speaking, French-speaking, and Italian-speaking Swiss.

The results of Gilbert, Shanmugam and Cavalleri (2022) corroborate those of Novembre et al. (2008). The genetic structure of Europe appears to be clinal but not homogeneous (Figure 12). The authors identify three main groups: North-West, Central-East, and South. Ashkenazi Jews differentiate within this continuum.

In the United Kingdom, when analyzing a sample of populations whose four grandparents are native to the same region, the genetic differences between micro-geographical regions are very clear (Leslie et al., 2015). Yet, the United Kingdom is an advanced country with modern transport modes. The mixing of populations is much less extensive than one might intuitively believe. Even when individuals are sampled randomly, without taking into account the birthplace of their grandparents, the stratification within the British people persists, but to a lesser extent (The UK10K Consortium, 2015).

It follows from these results that the European population is not genetically homogeneous and is *stratified*. This fact must be taken into account in applied studies, particularly those related to diseases.

Figure 12: Evolutionary tree showing genetic kinship degrees between different European populations. Links to geography are obvious (adapted from Gilbert et al., 2022).

Population Admixture: A Non-Specific Phenomenon to Europeans

Europeans run on admixture. But this trait is not original. Indeed, this is the case for the vast majority of current human populations. Classic physical anthropology had glimpsed this phenomenon and translated it according to the vocabulary of the time, with a distinction between "primary races" and "secondary races." The latter corresponded to the majority of modern populations and were supposed to result from repeated mixing (Hooton, 1926).

The three ancestral components of current Europeans are ancient, with the most antique, the original hunter-gatherers, dated to the Mesolithic (from – 15,000 years to – 5,000 years in Europe).

Much more recently, in the current era, the populations denominated in the United States as "Latinos" or "Hispanics" represent such an admixture, forged by contemporary history. These "Hispanics," inhabitants of current Latin America, have three main ancestral components: native Americans, i.e., the first inhabitants of the place, Europeans, mainly of Iberian origin (the conquistadors), and Africans, resulting from the slave trade. Depending on the origin, the percentage of these three components varies greatly. In Bolivia, the native American contribution is predominant, as this population still constitutes the majority of the current population of this Andean country today. In Ecuador, the native American component is stronger inland, while the African input is more significant on the coast.

Europeans are therefore a mixture. There is no "pure European race." But it is not just any mixture. Mixture does not mean homogenization. Current European populations share certain genetic characteristics that unambiguously distinguish them from the other major continental populations.

Blond Hair and Blue Eyes

Among European populations, the "Nordic" type individuals—blond hair and blue eyes, which are not found on any other continent—have not only made Nazi theorists ponder. They seem to arouse a certain fascination, which translates into an overrepresentation among actresses and actors and in advertising.

This phenotype, which is more frequent in Northern Europe, was decreed by the Nazis as the emblem of the "pure Aryan race." Adolf Hitler, who was himself dark-haired, and Heinrich Himmler, not exactly blond either, nurtured dreams of a future Germany populated by one hundred million blond Aryans. They seem to have been stunned and considerably disappointed to note that there were more blond people in Poland, a contemptible Slavic country, than in Germany. Of course, the disappointment was exacerbated when, at the Berlin Olympics in 1936, their "Aryan" champions were severely beaten by the African American champion Jesse Owens.

These anecdotes aside, the "Nordic" type appears as the ultimate expression of the "White" (Caucasian, European) race. What is the antiquity of this specific phenotype? Does it present a selective advantage, apart from the asset of making a career more easily on television or in cinema? White skin in general, especially if it is very fair, is linked to the metabolism of vitamin D, the antirachitic vitamin. Very fair-skinned people absorb UV more easily and thus synthesize provitamin D better in northern regions, where UV radiation is reduced, especially during winter (Cochran and Harpending, 2009).

An original hypothesis was proposed by Wilde et al. (2014): white skin, blue eyes, and blond hair may have been favored by the "curse of agriculture," the mother of all bad diets. The somewhat irenic and mythical notion that humans were happier and ate better when they were hunter-gatherers, a new avatar of the "lost paradise," has been ardently defended by Yuval Noah Harari in his successful book: "*Sapiens*: A Brief History of Humankind (2011)". More rigorously, Wilde et al. (2014) speculate that the hunter-gatherer diet was richer in vitamin D, with a high intake of meat and fish (especially fish liver), like current Inuits. The new diet following the Neolithic* agricultural revolution emphasized the consumption of cereals to the detriment of animal proteins. This would have paved the way for an indirect natural selection for lighter skin, to compensate for the vitamin D deficiency. According to this hypothesis, the selection favoring light skin would be quite recent, dating back to the Neolithic at most, i.e., a few millennia. The case of the remarkable and recent adaptation of the Aymara native Americans of Bolivia and Peru to altitude (see later: adaptation to altitude), which established itself in just a few millennia, shows that natural selection can be very fast. For skin and eye color, this speed would be facilitated by the fact that these traits seem to be governed by a limited number of genes. According to Sablonnière (2020), skin color would be dependent on about twenty genes, while a single gene would condition 90% of eye color. Two genes seem to have a preponderant influence on the genesis of the "light skin" phenotype: *SLC24A5* and *SLC45A2* (Cochran and Harpending, 2009; Crawford et al., 2017; Gibbons, 2015; Novembre and Han, 2012; Quintana-Murci, 2021; Sablonnière, 2020).

What Was the Skin and Eye Color of Paleo-Europeans?

The least one can say is that the data is contradictory.

According to the hypothesis of Wilde et al. (2014), our ancestors turned pale not very long ago, about ten millennia, with the advent of agriculture. But before? Europe has been populated for much longer. What was the skin color of Cro-Magnon and his congeners?

According to the biologist Evelyne Heyer (Heyer, 2025; Stricot, 2023), old Europeans had dark, even black skin, and possibly blue eyes. Following the hypothesis of Wilde et al. (2014), Heyer (2025) postulates that the artists who painted the Lascaux cave had black skin. The cause of the skin lightening of Europeans would be the carbohydrate-rich and protein-poor diet introduced by the advent of agriculture. This modified diet would have led to a vitamin D deficiency, and as a counter-effect, the "necessity" of white skin to better absorb solar UV, which generates provitamin D. A genetic contribution from Middle Eastern farmers would have accelerated this very recent whitening.

This hypothesis of dark-skinned paleo-Europeans seemed confirmed by the discovery of the famous "Cheddar Man," the father of all Britons, but who would have only very distantly resembled a City gentleman. Indeed, "the Cheddar man" was supposedly coal-black, but with blue eyes. This is unusual, because as a rule, selection for these two traits does not go hand in hand. The Cheddar man was discovered in 1903 in Cheddar, the village of origin of the famous cheese. He dates back to the Mesolithic era, i.e., about ten millennia. The Natural History Museum in London published a press release revealing the surprising physiognomy of the ancestor of all Britons, deduced from the morphology of his skull and the analysis of his DNA. However, this handsome dark-haired, blue-eyed man is certainly not the sole ancestor of the current inhabitants of Great Britain. It is thought that the genetic component of the Cheddar man's ancestral population constitutes at most 10% of the current British genome (Bordenave, 2018b).

Among human phenotypic traits, apart from cognitive characteristics, skin color is probably the one with the strongest political and ideological connotation. The skin color of our ancestors has therefore been massively used for ideological purposes to the detriment of balanced and rigorous science. It is probably for this reason that the Cheddar man's skin color benefited from such hasty and noisy publicity. Many debaters saw a perfect opportunity to deflate the "balloon of white arrogance." As these lines are being written, we are still awaiting the publication of these results in a reliable international scientific journal with a peer review committee: they have never gone beyond the "preprint" stage. The rapid circulation of unconfirmed preliminary scientific results in the media is regrettable (Bordenave, 2018b).

Fortunately, more reliable data are available. According to Gibbons (2015), paleo-Europeans who lived 8,500 years ago in Spain, Hungary, and Luxembourg had a dark complexion. In contrast, their contemporaries (–7,700 years ago) in southern Sweden were carriers of two genes, *SLC24A5* and *SLC45A2*, indicating light skin, and also the *HERC2/OCA2* gene, denoting blue eyes. According to this author, Near Eastern farmers who migrated to Western Europe during the Neolithic were carriers of the "light skin" genes, which they diffused into local populations. Still according to Gibbons (2015), light skin only became widespread among Europeans 8,000 years ago.

Much older *sapiens*, whose age is about 45,000 years, discovered in Germany and Czechia, lacked the genes conferring light skin (Curry, 2024; Sümer et al., 2024).

To complicate the picture further: during the long period of mixing between *sapiens* and Neanderthals (between 50,500 and 43,500 years ago, a period lasting about 7,000 years; Iasi et al., 2024), our archaic cousins may have gifted us (adaptive archaic introgression) with "light skin" genes, destined to be positively selected in weakly sunlit climates like that of late Pleistocene* Europe (Louvet, 2024a). Indeed, there are good reasons to think that our "archaic" cousins had light skin and eyes, and blond or red hair, like many current Northern Europeans (Condemi and Savatier, 2016).

In conclusion: what was the skin color of paleo-Europeans? It remains hypothetical. It seems likely that it was dark, at least in some populations and at certain times, but it is not certain that this was the case everywhere and at all times.

The Discreet Charm of Obsolete Classifications

In the 1960s, French high school students were still studiously learning in geography class that Europeans presented three main types: Nordic, tall blonds with blue eyes; Alpine, dark-haired with blue eyes; and Mediterranean, dark-haired with brown eyes and "olive" skin (see above: Classifications of the Modern Era (Since the Renaissance). As previously outlined (see "origin of human races"), this "classification" was very scientifically decreed by the American eugenic journalist Lothrop Stoddard. It has, of course, fallen into disuse, but ultimately, it is clear that people in Rotterdam and Naples do not look exactly the same, which is a pleasure for the eyes.

French: Their Ancestors the Gauls?

An important clarification: I am referring here to the so-called "native French", jokingly nicknamed "Gauls" by young people from the suburbs who come from more recent migratory flows. The very existence of our supposed ancestors, the "Gauls," is being questioned by quite a few contemporary authors. Let us see what the situation is.

The ANCESTRA project of the Jacques Monod Institute (https://anr.fr/Projet-ANR-15-CE27-0001) sheds informative light on those who preceded modern French on this land. It has confirmed that our Mesolithic hunter-gatherer ancestors received two major migratory waves: farmers from the East (Anatolia), 6,000 years BCE, and nomads from the Pontic-Caspian steppes north of the Black Sea (the Yamnaya, already mentioned), 3,000 years BCE. Apparently, the Yamnaya spread a male dominance, as evidenced by their Y chromosome, which is still dominant in Western Europe. This could suggest an elite that established itself (Geigl, 2020). A comparable phenomenon exists in Latin America, where the Y chromosome, inherited only from men, is predominantly European, in this case Iberian, while mitochondrial DNA, transmitted by women, has a larger Native American component. Geigl (2020) notes that between the Bronze Age and the Iron Age (the Gauls?), populations appear to have genetically stabilized and, remarkably, are not very different from current populations. This denotes a certain genetic stability of the French population since antiquity.

With the markers used for the study in question, the genetic differences between Gauls (a term used by Geigl, 2020), Germanic peoples, and Romans do not appear to be very significant, certainly less so than those observed between the original hunter-gatherers, the steppe migrants, and those from Anatolia.

Probably due to insufficient resolution of the genetic markers used (ancient DNA is often highly degraded), this study did not reveal a phenomenon in the paleo-populations under study that can be observed in modern populations, namely *genetic microstratification*. In the United Kingdom, as already noted, when analyzing a sample of populations whose four grandparents were native to the same region, the genetic differences between geographic microregions are very distinct (Leslie et al., 2015). However, the United Kingdom is an advanced country with modern transportation. Population mixing there is much less extensive than one might intuitively believe. Even when individuals are sampled at random, without considering the birthplace of their grandparents, stratification within the British population persists, but to a lesser extent (The UK10K Consortium, 2015).

In France, a very similar picture is found (Saint Pierre et al., 2020). Two independent studies covering all of mainland France and 2,184 contemporary individuals, selected only on the criterion of birthplace, yielded consistent results. A first split separates the South-West from the rest of the country. A second dividing line distinguishes six groups: South-West, South, Center, North-West, North, and South-East. There are differences between the populations north and south of the Loire, which roughly corresponds to the former domains of the langues d'Oïl and langues d'Oc (the former von Wartburg line).

I have outlined above who the ancestors of current Europeans were: Steppe peoples (Yamnaya), farmers from the East (Anatolia), and Mesolithic hunter-gatherers (Haak et al., 2015). Current French people are no exception to the rule. They harbor in their genes the three components common to all Europeans. This therefore dates back much further than "their ancestors the Gauls." Indeed, comparison with data from the 1000 Genomes Project (The 1000 Genomes Project Consortium, 2015), the Human Genome Diversity Project (HGDP; Cavalli-Sforza, 2005), and fossil DNA makes it possible to identify the three same founding populations in France as in the rest of Europe: (i) original hunter-gatherers; (ii) farmers from the East (Anatolia); (iii) Steppe peoples. Component (ii) constitutes 46.5–66.2% of the French population›s genome, with a North-South cline. Component (iii) is 19.6–41.2%. It is more significant in the North-West. Finally, the contribution of component (i) is 6.3–16.2%. It is highest in the South-West. In the extreme South-West, a Basque genetic contribution is noted, which aligns with linguistic data. The specific case of the Basque people will be discussed later. Geographical barriers (the rivers Garonne and Loire, mountains) are reflected in the genetic diversity of the French, whose population structure appears to be a miniature of the European population. This fine geographical structuring must be taken into account in studies associating rare or low-frequency genetic variants, which are important for medical research (see above: "Different strategies for analyzing the genetic control of diseases").

It is remarkable that the genetic variability of contemporary French people, far from being homogenized by recent mixing, continues to reflect structures that bear witness to a distant past.

In any case, the ancestors of the French do not seem to have been perfectly peaceful. According to Fernández-Crespo et al. (2025), evidence is found in North-East France of invaders from another region who would have been better off staying at home. They appear to have been mutilated and massacred, with the crime dating back to approximately 4,000 years before Christ.

Have the Indo-Europeans Been Found?

The Indo-Europeans are a burning topic, overloaded with ideology. Indeed, the concept was co-opted by the Nazis, who made the Indo-European conquerors the very symbol of the superiority of the "Aryan race," and contemporary Germans the purest descendants of this "Aryan race." Moreover, regarding this "superior race," the Nazis had merely followed in the footsteps of two French thinkers, George Vacher de Lapouge (a socialist, by the way) and Count Arthur de Gobineau, author of the famous treatise "Essay on the Inequality of the Human Races" (1853). It must be emphasized that these two authors were quite in agreement with the thought of their time. At that time, the superiority of the white man over other peoples was self-evident (to Europeans), with the white man preferably being of Anglo-Saxon origin. The specificity of Vacher de Lapouge and Gobineau was this focus on Nordism: the peoples of Northern Europe were seen as superior to all others, including Southern and Eastern Europeans. The Nazis recognized this affiliation with their "Gobineau ist unser" (Gobineau is ours). In a more polished register, Indo-Europeans are one of the favorite themes of the French contemporary school of thought "Nouvelle droite" (New Right) and their leader, Alain de Benoist. The emblematic magazine of the "Nouvelle droite," "Nouvelle École," (New School) immediately captured all the interest of ancient DNA studies for their favorite theme, and recently published a special issue on the "paleogenetics of the Indo-Europeans," which is well-documented, moreover (Bouts, 2019).

Initially, "Indo-Europeanism" is a purely linguistic concept. The hypothesis of a group of related languages unified under the term Indo-European languages was first proposed by the British philologist Sir William Jones, as early as 1786 (Cavalli-Sforza and Cavalli-Sforza, 1997). Sir William was a judge in India, which was then under British rule. Studying Sanskrit, he was struck by the resemblance of India's sacred language to numerous European languages (Harari, 2011). The French anthropologist Georges Dumézil (1898–1986), extensively cited by authors of the "Nouvelle droite," considerably enriched this theme of the Indo-Europeans, not only through his work on linguistics but also on mythologies, religion, and the structure of Indo-European societies. The highly controversial "Dumézilian trifunctionality" is supposedly found in all peoples of Indo-European descent, both in India and in Europe. It dictates that Indo-European societies are structured around the sacred, religious function, the warrior function, and the function of production (farmers). One of the preferred themes of the "Nouvelle droite" is that this trifunctionality has persisted among modern European peoples. Currently, it is estimated that about half of the world's population speaks one of the roughly 400 cataloged Indo-European languages, in Europe, Central and South Asia, and Iran (Callaway, 2025).

Well beyond this purely cultural debate, the current controversies concern the physical, biological existence of the Indo-Europeans. The idea of Lapouge, Gobineau, and the Nazis was that Northern Europeans, the "Aryans," were the direct descendants of the Indo-European conquerors. The Nazis placed the origin of these conquerors in present-day Germany. A diametrically opposed thesis is that of the French archaeologist and prehistorian Jean-Paul Demoule (2014), who defends the idea that Indo-European languages spread, not through conquerors, but through purely cultural contacts. For this author, "that an ancestral people disseminated their language from a single cradle is a reductive and simplistic model" (cited by Foucart, 2009).

However, recent genetic studies seem to support the physical migration hypothesis rather than that of an exclusively cultural diffusion, to the point that the journal *Nature* (Editorial, 2018) expressed concern about the danger that such results could be co-opted by extremists to revive the theories of the German historian Gustaf Kossinna (1858–1931). This author proposed a theory according to which a regionally circumscribed ethnicity can be characterized by the archaeological industry found at a given site (a theory called culture/history archaeology, or simply, culture/history theory). For him, the "Corded Ware culture" industry was associated with Indo-European peoples and their migrations. According to Callaway (2018), the results of Haak et al. (2015) seem to support Kossinna›s theory. Haak et al. (2015) analyzed more than 2,500 individuals belonging to 203 different populations using 400,000 SNPs. As I have already explained, they postulated that current European populations are a combination of three founding populations: (i) Mesolithic hunter-gatherers; (ii) farmers from the East (Anatolia); and (iii) Steppe pastoralists, Yamnaya (Mathieson et al., 2015). It has been proposed that these Yamnaya pastoralists are at the origin of the spread of Indo-European languages and the Corded Ware industry (Novembre, 2015).

At least in Southern Denmark, the advance of the Indo-Europeans (Steppe people) appears to have been violent, with the near elimination of earlier populations (Allentoft et al., 2024; Leroyer, 2024).

The data from Haak et al. (2015) are more in favor of the "Steppe hypothesis" (according to which Indo-European languages originated from peoples coming from present-day Ukraine and Russia) than the "Anatolian hypothesis." However, the latter retains its supporters (Balter, 2015), and the debate recently resurfaced (Callaway, 2025). This linguistic debate concerns the origin of Indo-European languages, and "Proto-Indo-European" (PIE; Balter, 2015). On the other hand, the fact of a "massive" demographic migration of the "Steppe peoples" to Europe some 4,500 years ago (Haak et al., 2015) has not been refuted.

Finally, according to Novembre (2015), the Steppe pastoralists may originate from Northern Eurasian populations and may have contributed to the genetic makeup of ancient North American Indigenous populations. A recent study (Pankratov et al., 2024) indicates that this "Steppe people" component influences several phenotypic properties of modern European populations.

In a recent study, the team of David Reich, the "pope of ancient DNA," cuts straight to the chase, abandoning obscure appellations (Yamnaya, Steppe peoples) to speak plainly about the genetic origin of the *Indo-Europeans* (emphasis added by me) (Lazaridis et al., 2025). According to this study, based on the analysis of ancient DNA from 435 individuals, there were three populations at the origin of the Yamnaya, the most important of which was the "Caucasus-Lower Volga cline." The final unity of the speakers of "Proto-Indo-Anatolian," the ancestral language of the Anatolian and Indo-European peoples, would have occurred between 4,400- and 4,000-years BCE. According to the study by Lazaridis et al. (2025), analyzed by Callaway (2025), pre-Indo-European languages would have originated in the Caucasus and the Lower Volga, followed by a double migration towards Anatolia and towards the Black Sea steppes, from where the Yamnaya would have disseminated these languages westwards and eastwards. Bronze Age Anatolians would have spoken Hittite, the first written Indo-European language (see also the analysis by Bordenave, 2025). Linguistic hypotheses remain speculative. In contrast, genetics can strongly support the reality of migratory routes. This is, moreover, one of the most robust aspects of the hypotheses discussed in this book: population genetics and phylogenetics make it possible to reliably trace the genealogies, migrations, and degrees of kinship of human populations.

It thus seems that the Indo-Europeans have indeed been found. But bad luck for the Nazis: the origin of the Indo-European conquerors was, not Germany, but Ukraine and Russia, i.e., the territory of the Slavic peoples, considered inferior by the theorists of the 3rd Reich.

The Population Genetics of India: A Persistent Racialized Social System

Studies (Narasimhan et al., 2019; Schaefer and Shapiro, 2019; Wade, 2018) have examined the origin of the current populations of the Indian subcontinent. According to these studies, based on the analysis of ancient DNA, although current India is divided into many ethnic, linguistic, and religious subgroups, all these populations originate from three Eurasian ancestral groups: (i) local hunter-gatherers; (ii) farmers from the Middle East; and (iii) Steppe pastoralists (the "Yamnaya culture," already mentioned above). This picture is perfectly symmetrical with that of European populations, with their three founding populations. According to Schaefer and Shapiro (2019), the "Steppe peoples" arrived in India later than in Europe, around 1,900–1,500 years BCE.

According to Quintana-Murci (2021), there is a North-South gradient in India for these components, with the Yamnaya contribution decreasing towards the South.

These studies seem to confirm that the Yamnaya Steppe pastoralists, who migrated westwards as well as eastwards, are indeed the origin of the Indo-European languages. This would explain the kinship of the latter with Sanskrit, the sacred language of India, and would confirm the premonitory intuition of Judge Sir William Jones.

It should be noted that these Eastern "Indo-Europeans" are the subject of the same political co-option as their Western cousins. In this instance, it is current Indian nationalists who are rewriting history by emphasizing the predominance of Hindu ancestry in modern India (Hindus constitute 80% of the population, compared to only 14% Muslims), with Hindus supposedly of "Indo-Aryan" lineage (Viguier, 2023). However, India is a decentralized federal country, and not all regions adhere to this narrative, conveyed by school history textbooks. South Indians pride themselves on being of "Dravidian" rather than Aryan descent (see below).

According to Dobzhansky (1962a), the ancient masters of India believed in the hereditary transmission of psychological traits, hence the establishment of a system of endogamous (reproducing only within themselves) castes based on professional specialization. The highest caste, the Brahmins, consider themselves to be the direct descendants of the Indo-European conquerors who subjugated the local aboriginals and established this caste system, which is more than 2,000 years old. The Parsi caste, for its part, is of Persian origin (Dobzhansky, 1962a).

According to Reich (2018c), two caste systems coexist in India. The "Varna" system is composed of the following castes, from top to bottom: (1) Brahmins and warriors; (2) merchants, farmers, and artisans; (3) workers; (4) untouchables. This evokes the divisions of Indo-European society theorized by Georges Dumézil. Independently of these "official" categories, the "scheduled tribes" correspond to unassimilated groups, probably descended from the original aboriginals of India.

In Malaysia, Thailand, and Laos, such original "tribes" persist, called "Mlabris" or "Manis." In Thailand, these people are nicknamed "forest rambutans." This alludes to their frizzy hair like that of Africans, the rambutan being a fruit covered with wavy hairs.

The "Jati" system is much more complex than the "Varna" system and is thought to include up to 4,000 endogamous groups. According to Reich (2018c), the average genetic differences between the castes of the "Varna" system are 3 times greater than those observed between Northern and Southern Europeans. This millennia-old endogamous system has therefore produced measurable results.

The caste system is officially abolished in modern India. However, there is no doubt that it largely persists in people's minds and continues to be reflected in the socio-economic scale.

According to Quintana-Murci (2021), men were more assigned to their caste than women, who could "move up" to higher castes. This is reflected in the polymorphism of mitochondrial DNA, which is exclusively maternally transmitted.

The Mysterious Basque People

The Basques constitute a very specific population living in an area located in the South-West of France and the North-West of Spain. They consider themselves to be very different from their French and Spanish compatriots. This sense of identity is undoubtedly more marked among the Spanish Basques, to the point that there was a long and bloody war of independence among the latter. The Basques speak one of the rare languages in Europe that is not of Indo-European origin.

The tradition among these proud people tells that they have lived in this territory for a very long time and have very ancient roots. Ancient DNA suggests that this is not a legend. Günther et al. (2015) sequenced the DNA of eight individuals from the El Portalón ("the great gate") cave in the Spanish Basque Country. These individuals lived 3,500 to 5,500 years ago. Günther et al. (2015) also analyzed the genetic composition of contemporary Basques and those of other Iberian and European populations. In comparison with other Iberian populations, current Basques show fewer components from the Caucasus, Central Asia, and North Africa. Furthermore, they are closer to the eight individuals from the El Portalón cave than to other contemporary Iberians. It is possible that the eight individuals from El Portalón correspond to ancient Basque farmers and are the direct ancestors of the current Basques.

A Burning Topic: The Biological Specificity of Jews

Jews have been the object of fierce persecution until the recent past. The Nazi regime focused on the supposed nuisance of the Jewish "race" to promote its extermination policy. The Vichy government in France enacted anti-Semitic laws aimed at prohibiting Jews from many key jobs, with the goal of reducing their "harmful influence." Several propaganda exhibitions were organized in France so that the French could identify individuals "of Jewish blood." To some extent, the French leaders of the time went further than the Nazis. Indeed, unlike the Vichy government, the latter did not consider Jews to be a race in the biological sense of the term, but rather a "community of spirit" (Rindermann, 2022). However, they did not consider Jews to be white (Taguieff, 2020).

The controversial Georges Vacher de Lapouge, already mentioned, had once and for all established a hierarchy among the peoples of Europe. At the very top were the Aryans, of Germanic descent, tall, blond, dolichocephalic (i.e., with an elongated skull viewed from above), born to command. At the bottom were the brown brachycephalics (the opposite of dolichocephaly), born to be docile and execute. By a hardly conceivable historical anomaly, these brown brachycephalics took power during the French Revolution. In this *hit-parade*, Jews were considered intelligent, but carriers of all defects, incapable of creating, and doomed to profit from the work of others, particularly the docile brown brachycephalics (Ruffié, 1983).

This torrent of ostracism and this tragic past explain why discussing the biological specificity of current Jews is so delicate. Some authors have even asserted that Jews had no biological specificity and constituted a purely cultural and religious entity (L'Herminier and Solignac, 2005; Ruffié, 1983).

However, perhaps with nationalist ulterior motives, Israeli science has conducted numerous studies on this topic. Scientists whom no one would dream of accusing of racism, such as Professor Cavalli-Sforza, an undisputed moral authority, have conducted pioneering work on this topic (Carmelli and Cavalli-Sforza, 1979). In my opinion, seeking to know whether Jews, due to their very specific origin and history, have distinct genetic characteristics from those of neighboring populations should not constitute a taboo subject. However, this seems to be the case. Now, everyone has ancestors and depending on the population considered and its past history, they are not exactly the same ancestors. Modern high-resolution genetic markers seem to confirm that individuals of Jewish descent can be identified at the population level, and even at the individual level (Cochran and Harpending, 2009; Guha et al., 2012; Need et al., 2009). These analyses make it possible to distinguish Ashkenazi Jews (a Jewish population primarily originating from Central Europe; "Ashkenazi" is the Hebrew term for "German") from other Europeans. This thus confirms the pioneering work of Carmelli and Cavalli-Sforza (1979), conducted with genetic tracers of much lower resolution than the latest genetic tools.

Guha et al. (2012) compared Ashkenazi Jews with populations from Europe, the Middle East, and Central and South Asia. Ashkenazi Jews appeared to be intermediate between Europeans and Middle Easterners. This indicates that current Jews have retained a not insignificant component of their Middle Eastern origin, despite long centuries of coexistence with other European populations. This rather surprising result can be explained by a propensity for strong endogamy, due partly to persecution and partly to cultural and religious traditions.

In the study by Banda et al. (2015), covering a sample from Northern California, Ashkenazi Jews were identified by questionnaires on religion and nationality. This work confirmed that this human group shows certain genetic characteristics. The same applies to the study by Gilbert, Shanmugam and Cavalleri (2022) focusing on European populations.

These hypotheses have been challenged by Israeli geneticist Eran Elhaïk, from the University of Sheffield (United Kingdom), according to whom Ashkenazi Jews are massively of Caucasian Asian origin, and not of the Near East (Mary, 2015).

The considerable efforts made by many authors to increase our knowledge of the population genetics of contemporary Jewish populations have produced numerous medical benefits. Populations of Ashkenazi Jewish descent suffer from several genetic diseases with a greater frequency than other human groups (Cochran and Harpending, 2009). A detailed population genetics framework optimizes the genetic mapping of these diseases (Guha et al., 2012).

An original but highly controversial hypothesis already presented (see: "Frequent neurodegenerative diseases in Ashkenazi Jews") has been proposed to explain the undisputed fact that the average IQ (Intelligence Quotient) of Ashkenazi Jews is higher than that of other Europeans: 115 versus 100. This fact was by no means denied by the English scientist Francis Galton, one of the fathers of eugenics, who was nonetheless a notorious anti-Semite.

A Passing Note on Science and Ethics

"Nature is neither moral nor immoral. It is radiantly, gloriously, amoral." (Théodore Monod, French biologist and explorer)

The examples of the Indo-Europeans and the Jews serve to illustrate an important fact, which is one of the central themes of this book: scientific results disregard ideologies, whatever they may be, whether right-wing or left-wing. As the explorer Théodore Monod so accurately wrote: "Nature is neither moral nor immoral, it is radiantly, gloriously, amoral." It is therefore risky to try to base an ideology on natural facts, and even more so to attempt to twist scientific results to fit moral ideas, however noble they may be. Based on current findings, the Indo-Europeans do seem to correspond to a physical, biological reality, which satisfied the "New Right" ideologues. However, this people's cradle was not Germany, but Russia and Ukraine, much to the chagrin of the Nazis, had they known. Similarly, the fact that Ashkenazi Jews, on average, have a higher IQ than other Europeans is nothing that would satisfy anti-Semites. I will return to these considerations at length in Part II of the book (ethics and politics).

Paleo-humans: Species or Subspecies (= Geographical "Races")?

Bringing up our distant ancestors is not out of place in the context of the debate on the biological status of geographic populations. The problem has been largely discussed in previous works (Tibayrenc and Ayala, 2017b, 2020). Only the points specifically relevant to the present book will be set out.

Our ancestors pose a taxonomic (relating to classification) conundrum, which is that of paleo-species, in other words, extinct species, for which no clear definition is available (Langaney, 1999). I have already discussed the problem of the species in general, which is far from simple (see "biological and taxonomic reminder"). At least, the description of current species can rely on the criterion of interfertility (the "biological" or "mixiological" criterion of the species; Dobzhansky, 1937). To put it briefly: the donkey and the horse are different species because their offspring, mules and hinnies, are sterile—thus, there is an interruption of gene flow. In practice, many species continue to be described on purely morphological bases ("phenotypic species"), and gene flow is, in fact, not interrupted between certain different "species" (the polar bear, *Ursus maritimus*, and the brown bear, *Ursus arctos*, and their "pizzly" offspring). Thus, this criterion is often found wanting or simply ignored, but it has the merit of existing. For paleo-species, particularly those concerning our ancestors, it is much more complicated. Ancient DNA has changed the game, showing, for example, that *Homo sapiens* and *Homo neanderthalensis* had exchanged genes (Gibbons, 2017a), and possibly repeatedly. The data on this are contradictory. Either there were multiple episodes of gene exchange between *sapiens* and Neanderthals (Callaway, 2021), or there was a single, but very prolonged, episode (Bergström et al., 2020), up to 6000 years (Iasi et al., 2024; Sümer et al., 2024). But what is certain is that there has been interbreeding. So, are they species, or subspecies, or even (geographical) "races"? We are entering completely into the subjective and guesswork here.

It is therefore not surprising that the taxonomic status of our ancestors has led to incessant expert quarrels. This has generated a profusion of species names, all exacerbated by the weight of history, tradition, and also the egos of specialists. An exemplary case of this confused situation is that of the fossil discovered in Morocco in the 1960s at Jebel Irhoud. It was recently redated as 300,000 years old and has now been established as the oldest specimen of *H. sapiens* (Hublin et al., 2017). However, the same fossil had previously been identified as a Neanderthal upon its discovery, then possibly as *H. heidelbergensis* in 2007 (Gibbons, 2017b). The name *Homo helmei* was also proposed for this specimen (Scerri et al., 2018). This illustrates the subjective nature and uncertainty of paleo-species definition. Struck myself by the quite primitive appearance of the Moroccan *Homo sapiens*, I compared the cranial image communicated by Gibbons (2017b) with that of a Neanderthal and that of a modern human. The verdict from the ChatGPT oracle was unequivocal: "Skull 1 [Jebel Irhoud] shows similarities to Skull 2 [Neanderthal] due to its more elongated shape and the slope of the cranial vault, typical of ancient or intermediate hominids. In contrast, Skull 3 [modern human], with its modern profile, shows a more globular braincase and a smaller face, characteristic of *current Homo sapiens*. Conclusion: *The profile of Skull 1 is most similar to the profile of Skull 2* (my emphasis)." Even artificial intelligence has difficulty navigating this.

The blurry boundaries between *Homo heidelbergensis*, "pre-Neanderthals," and Neanderthals (Condemi and Savatier, 2016) are another illustration of the difficulty of describing and, above all, delineating paleo-species. This difficulty largely results from the gradual nature of the evolutionary process.

This digression on our ancestors aimed to once again expose the difficulty of taxonomic work: the debate about the status of human geographical populations, the subject of this book, is in no way an isolated case, as some defenders of "scientific antiracism" claim (see: Human races are difficult or impossible to define from a biological point of view). In all taxonomic work concerning living things, there is a degree of arbitrariness, subjectivity, and tradition. This does not mean that classifications of living things are a Figment of the imagination, quite the opposite. They are highly necessary for defining units of study and, generally, reliably reflect the degrees of genetic relatedness between living things. However, one must be fully aware of the methodological difficulties and uncertainties inherent in any taxonomic work.

Paleo-Racism vs. Paleo-Antiracism

It is interesting to note that prehistoric science has reflected, and still reflects, the ideological presuppositions of successive eras, proof, if any were needed, that the necessary "reciprocal sanctuarization" between science and ideology (Tibayrenc, 2019) is a very difficult and very necessary exercise. Our unfortunate cousin, Neanderthal man, has amply paid the price, for the worse, and now, for the better. In short, Neanderthal man has been politically "reclaimed."

The masterful description of the La Chapelle-aux-Saints man (Corrèze, France) by Marcelin Boule (1920) betrayed the prejudices of the time (see "who are our ancestors"), according to which the hierarchy of races was self-evident. This was compounded by the fact that the La Chapelle-aux-Saints man was an elderly subject afflicted with severe arthritis. Boule›s vision of the Neanderthal long struck the popular imagination and continues to do so: an extremely stocky, hunched man, with arms hanging like a gorilla›s, barely able to stand upright, readily hairy like an ape, and probably very mentally backward compared to modern man.

Stocky he was, reflecting an adaptation to the cold climates that our ancestor endured during his long stay in Europe. But the picture painted of him now is the opposite of that from the quite intolerant era of Marcellin Boule, a racist and colonialist period. Nowadays, emphasis is placed, conversely, on the fact that Neanderthal had a larger cranial volume than ours, that he piously buried his dead, as evidenced by the sophisticated burial of the La Chapelle-aux-Saints man, that he had produced graphic art (Louvet, 2024b), and that, all things considered, he was quite similar to us, as evidenced by the cover of the very attractive book: Néandertal, mon frère ("Neanderthal, My Brother"), (Condemi and Savatier, 2016), on which our Paleolithic cousin appears as a kind of youthful Anthony Quinn in shirtsleeves. In short, all this reflects the current era, preaching tolerance and non-discrimination.

Recent Evolution in Humans

This is an important question in the debate about human races. Indeed, any biological differences observed between geographical populations must be the result of a relatively recent, differentiated evolution, subsequent to the "Out of Africa" event, i.e., a few tens of millennia ago.

Has Evolution Stopped?

A very widespread idea is that human biological evolution stopped some 40 to 50,000 years ago, and that the evolution of our species is now purely cultural. The reason invoked is that humans are now removed from natural selection, due to the advent of modern comfort and medicine. This view is championed by Gould (2000), Lawler (1978), Mayr (1963), and Ruffié (1983). The underlying idea is that the "Out of Africa" event is so recent (about 50,000 years ago) that "evolution has not had the time" to act. This seems erroneous. Evolution can be very rapid, as we will see with the cases discussed below, particularly adaptation to altitude.

Numerous Signs of Recent and Even Current Evolution

The idea of human evolution having still been active in the recent past, or even continuing into the present day, constituted the central argument of two recent works, which were subject to vehement criticism. These are: "The 10,000 Year Explosion" (Cochran and Harpending, 2009) and "A Troublesome Inheritance" (Wade, 2014). Nicholas Wade is a highly controversial British science journalist, who is not new to polemics. However, his book is rich in scientific information. There is no doubt that he has thoroughly researched his topic. As for "The 10,000 Year Explosion", it is the work of two scientists whose competence is indisputable, the second of whom, the late Henry Harpending, was a member of the United States National Academy of Sciences. Like Nicholas Wade's, their book is a goldmine of scientific information. Both works have been subject to virulent attacks, not just because they defended the idea of recent human evolution, but especially because they claimed that this evolution had been geographically differentiated, and that it particularly concerned cognitive characteristics—"potentially explosive" hypotheses (Richardson, 2011).

If there is recent, or even contemporary, evolution, it is logical to think that it is geographically differentiated: across the globe, ecotypes, environments, climates, and pathogens differ considerably. What do the data say?

Cultural evolution obviously plays a major role in human societies. However, recent and abundant data, drawn particularly from fossil DNA, the analysis of skeletons of vanished populations, historical data, and extensive genetic analyses, particularly from the UK Biobank (Gibbons, 2019), strongly suggest that biological evolution continues to leave its mark on current populations.

We have already seen the case of infectious diseases, particularly malaria. Since this disease strikes young children, only the survivors of the infection will be able to transmit their genes to their offspring. This is the very principle of natural selection. It is a classic notion in tropical medicine (Gentillini et al., 2012) that African children, even if they pay a heavy toll to malaria, are more resistant to it than European children.

Aside from infectious diseases, many characteristics still seem to be subject to natural selection. We have already mentioned (see: "what are Europeans made of?") the problem of the eye and hair color of our ancestors. Many other phenotypic traits appear to have undergone, or are still undergoing, evolution today.

Alan Cooper, from the University of Adelaide (Australia), has identified 57 genomic regions showing signs of selection subsequent to the "Out of Africa" event. The traits concerned include fat storage, neuronal development, skin physiology, and major current diseases such as metabolic syndrome and neurodegenerative pathologies (Mangin, 2023).

According to Murray (2020), evolution has been active since the Out of Africa event and has been geographically differentiated.

According to Sablonnière (2020), the height of all global populations has been increasing for the last two centuries. Changes in lifestyle are not enough to explain this fact. This change has been attributed to better nutrition. However, it is not certain that the diet of current adolescents is particularly healthy. Even in Thailand, a country renowned for its delicious and healthy "street food," teenagers are regulars at McDonald's and KFC. And yet, in the "country of smiles," the younger generations are also taller than their elders. According to Austad and Finch (2017), human height began to increase 500,000 years ago, then became 10% greater than the current average height. It then decreased 50,000 years ago, accelerating this decline with the advent of agriculture, which is attributable to a diet lower in protein (Harari, 2011). It seems that height began to increase again very recently, 150 years ago. In 1950, Americans of European descent were taller than the Dutch. Currently, the opposite is true. The Dutch are considered to be the people with the greatest height in the world.

According to Zampieri (2017), the age of first menstruation (menarche) is becoming increasingly later, which has an obvious impact on reproductive fitness.

Field et al. (2016) analyzed the allele frequencies of 3,195 British individuals. They deduced that current Britons have better lactose tolerance than their ancestors of 20 centuries ago, are blonder, have bluer eyes, and are taller.

A study (Marshall, 2025; Staeger, 2025) involving 8,000 women across three different countries (Australia, Mexico, Poland) shows that pelvis width has been decreasing for 150 years, at a rate of 0.42–0.47 mm per year. One possible cause invoked is a decrease in selection pressure for a wide pelvis occasioned by the proliferation of C-sections. This anatomical feature is the site of a "dilemma" between two opposing selection pressures: a problem-free birth requires a sufficiently sized pelvis, but the mechanical constraints inherent in bipedalism work in the opposite direction.

Radier (2020) notes that since the advent of agriculture, one can observe a gracilization of the body and a decrease in brain size. These facts are confirmed by Rühli and Henneberg (2017), who also note a higher frequency of periodontitis, cavities, and wisdom teeth. According to the same authors, brachycephaly (skull wider than long when viewed from above) has increased very recently (800 years), particularly in Central Europe. Finally, a phenomenon of "hypermorphosis" might be at work concerning height. Its increase may have been excessive, generating vascular and mechanical problems. According to Rühli and Henneberg (2017), gene flow primarily occurs over short distances and remains limited and slow. This, combined with different environmental pressures, leads to physiological and morphological differentiation between different geographical populations. The few specific cases presented below are indeed illustrative of geographical differences in the recent evolution of our species.

Adaptation to Altitude

This environment is characterized by aggressive sunlight, as the filtering of ultraviolet rays is less effective in a rarefied atmosphere, and by a decreased oxygen content. Three populations show remarkable adaptation to this inhospitable environment: Tibetans, Ethiopians, and the Aymara Indians of the Bolivian and Peruvian Altiplano. These three peoples have developed different mechanisms to adapt to altitude. This is a phenomenon of "convergent evolution," like the fins of fish and dolphins, or the wings of birds and bats. In these two examples, the possession of organs with similar functions is not due to genetic relatedness but to adaptations acquired in parallel.

Tibetans have specific mutations in the *EPAS*1 and *EGLN*1 genes, which are associated with oxygen metabolism (Gurdasani et al., 2019). It appears that these mutations result from archaic Denisovan adaptive introgression (Racimo et al., 2015). Tibetans show a significant difference for the *EPAS*1 polymorphism compared to Han Chinese, the dominant ethnicity in China. This specific *EPAS*1 gene polymorphism is found in other Himalayan populations: Chinese, Indians, Bhutanese, Nepalese (Simonti, 2025). It thus appears that all these populations received the archaic Denisovan adaptive introgression which favored their adaptation to altitude.

Having worked for over 7 years in Bolivia, I am well acquainted with the case of the Aymara Indians, who constitute the original inhabitants of the Altiplano, a high plateau located at an altitude of about 4,000 meters (13,123 ft). I can testify that life at this altitude subjects you to a very specific stress. First-time arrivals in this region frequently suffer from "soroche," or altitude sickness: violent headaches, nausea, vomiting. The cause is cerebral edema. Even after adapting, on a daily basis, shortness of breath is rapid after any physical exertion.

The traveler fortunate enough to visit the magnificent Lake Titicaca region will encounter people with a very particular physiognomy: short, stocky, with a very copper complexion (Figure 13).

Figure 13: A smiling young Aymara woman in traditional dress and her baby. The mother and child have a very copper-colored skin, and display the "Altiplano cheeks," which result from physiological polycythemia (high red blood cell count). This abundance of red blood cells is one of the mechanisms developed by the Aymara to adapt to the altitude. Indeed, it allows them to better capture rarefied oxygen.

The masterful studies of the "Instituto Boliviano de Biologica de Altura" (Bolivian Institute of High Altitude Biology; IBBA; https://ibba.umsa.bo/), a French-Bolivian institute where I worked for 7 years, have highlighted several informative facts about the adaptation of Altiplano populations to altitude. The absence of stable human settlements above 5,000 meters (16,404.2 feet) has been noted. This appears to be a physiological limit for our species. Among the populations of Bolivia, a very diversified country ("estado plurinacional de Bolivia"), the Aymara are by far the best adapted to altitude. They exhibit not only a physiological adaptation (the polycythemia shown in Figure 13) but also a metabolic and anatomical one: they have an increased lung capacity, evidenced by their "barrel-shaped" chest. It is remarkable that this adaptation was acquired in a relatively short period of time: it is estimated that the Aymara colonized the Altiplano some 5,000 years ago. This shows that human evolution can be very rapid in the event of strong selection pressure and dismantles the argument that the time elapsed since the "Out of Africa" event (some 50 to 60,000 years ago) is too short to have allowed substantial evolution among human populations.

As I already mentioned, altitude causes significant stress. During our stay in Bolivia, from 1980 to 1984, my wife underwent three miscarriages. The Bolivian doctors following her were adamant: this was due to the altitude, and this adverse phenomenon was frequent among European women. Once we returned to a low altitude, California in this case, there were no more miscarriages. This phenomenon was endured by the Spanish colonists during the time of the conquistadors. La Paz was initially founded on the Altiplano, at the current site of Llaja. Spanish women experienced the same obstetric problems as my wife. The colonists then decided to migrate to the current site of La Paz, which is a deep and wide canyon, descending to an altitude of 3,400 meters (11,155 feet), and relatively protected from the rigors of the climate. Babies that are not born will never transmit their genes: this is the very definition of natural selection. Aymara women do not experience these obstetric problems.

In addition to altitude adaptation, the Aymara's very copper-colored skin constitutes an effective protection against the cruel solar radiation (because UV rays are imperfectly filtered by the thin atmosphere) of the Altiplano.

In Bolivia, another important Native American population consists of the Quechua, who are the descendants of the Incas. They have therefore only been there for a few hundred years. Logically, and this has been verified by the IBBA, their adaptation to altitude is poorer than that of the Aymara. The Quechua have, moreover, tended to settle in mid-altitude regions (around 2,500 meters; 8202.1 ft), where the cities of Cochabamba and Sucre are located.

Mestizos ("Chollos") are an important minority in Bolivia. They constitute the bulk of the middle class in large cities. According to the IBBA, their adaptation to altitude is less than that of the native populations.

Finally, the rare Bolivians of "pure" European descent can experience serious health problems. Even among those born in La Paz, a pathological polycythemia can emerge, forcing them to migrate to the lowlands, to Santa Cruz de la Sierra.

It is therefore entirely erroneous to speak of a "modest" adaptation to altitude in humans, and to attribute it to acquired and reversible, non-genetic characteristics

(Ruffié, 1983). Similarly, in light of the data concerning the Aymara, we can reject the assertion by Omiye et al. (2023) that differences in lung capacity between geographical groups inferred by artificial intelligence are artifactual.

Lactose Tolerance

Lactose is the sugar in milk. An enzyme called lactase allows for its digestion. Infants logically can metabolize lactose, but this lactose tolerance tends to disappear with age in most of the world's populations. However, some populations continue to digest their coffee with milk without problems. Lactose tolerance is another highly illustrative example of recent evolution, as well as a manifestation of the "Baldwin effect," which refers to the impact of learned behaviors on natural selection (Richerson et al., 2010). The hypothesis put forward is that European populations, particularly those in the north of the continent, began consuming milk and dairy products a long time ago, which would have led to the selection of genes permitting lactose tolerance (Zampieri, 2017). The distribution of lactose tolerance is unevenly spread across Europe (Sablonnière, 2020). It is believed to be present in 82% of Swedes, but only in 43% of French and 6% of Italians (Henry, 2019).

It is obviously difficult to know whether Northern Europeans began drinking milk because they were lactose tolerant or acquired lactose tolerance because they drank milk. The second hypothesis is favored by Lala and Feldman (2024). It is the eternal story of the chicken and the egg. This lactose tolerance seems to have appeared a few thousand years ago (Attwaters, 2024; Cochran and Harpending, 2009; Quintana-Murci, 2021). According to Gibbons (2015), neither Paleolithic European hunter-gatherers, nor Near Eastern farmers who came to Western Europe, nor the steppe pastoralists who arrived later were lactose tolerant. This trait would only have spread in Europe 4,300 years ago.

The study by Field et al. (2016) shows more precisely that the selection for lactose tolerance operated very recently, over the last two millennia. American "identitarians" have made lactose tolerance an emblem of the "white race" with public milk consumption, a process, incidentally, tinged with humor and self-mockery. Indeed, as already mentioned, this tolerance is absent in most global populations. But the posturing of American "identitarians" is all the inepter because certain African peoples (Reich, 2018c): the Fulani in West Africa and the Maasai in East Africa, who are also milk drinkers, also possess this lactose tolerance. According to Quintana-Murci (cited by Dos Santos and Levino, 2021), Pygmies are also lactose tolerant and would have gifted this trait to their Bantu neighbors following interbreeding. In a swap of good deeds, the Bantu offered the Pygmies the gene for sickle cell anemia which, as we have seen, confers good resistance to malaria on heterozygous carriers. The virtues of interbreeding.

Elsewhere in the world, after weaning, babies stop easily digesting milk. According to Cavalli-Sforza and Cavalli-Sforza (1997), yogurt and cheese are not concerned, as the lactose they contain is predigested by the microorganisms involved in their making.

Starch Digestion

By transitioning from a hunter-gatherer lifestyle to a farming one, populations changed their diet: less meat products, more carbohydrates, particularly starch. They adapted, relatively recently (within the last 12,000 years), through duplication/deletion of the amylase gene, the enzyme that converts starch into glucose. The mode of evolution of this gene varies greatly from one continental population to another (Bolognini et al., 2024).

Recent Evolution of Cognitive Traits

This is one of the main, considerably incendiary themes of the books by Wade (2014) and Cochran and Harpending (2009). For these authors, recent evolution is far from having concerned only the age at which young girls had their first menstruation, or how infants digested milk. It would have also, and above all, acted on cognitive abilities, and this, *differently from one region of the globe to another*. Such ideas are truly likely to have "potentially explosive implications" (Richardson, 2011). To expose them would be "playing with dynamite" (Molteni, 2022).

By evoking the possibility of cognitive differences between geographical populations, we undoubtedly touch on the very essence of racism. This simple idea appears quite simply scandalous to many. According to Winegard, Winegard and Anomaly (2020), for many authors, it is a question of an ethical principle to reject the "hereditarian theory, "which dictates that at least part of cognitive faculties is of genetic origin. We read correctly: in the name of ethics, a scientific theory is rejected *a priori*. This is the major cognitive bias that I am trying to highlight in this book. A number of authors, to maintain this preeminence of ethics over scientific rigor, do not hesitate to resort to "noble lies" (Winegard et al., 2020)." But lies may be noble, they remain lies, which is not good for scientific accuracy. Alexandre (2018) unambiguously supports this strategy: "exceptionally, scholars must put scientific truth after the fundamental philosophical principle of the equality of all groups of men."

However, as I emphasize throughout this book, ideology, indignation, and more generally, emotion, are bad counselors in science. One can even say that they are the number one enemies of rigorous scientific reasoning. It is useful to recall here that many scientific theories appeared scandalous in their time while being well accepted today (Tibayrenc, 2019). Furthermore, it cannot be overstated that differences do not mean hierarchical ranking (Tibayrenc, 2022). Camembert and Belgian chocolate do not have the same taste. Can we say for all that that Belgian chocolate is "better" than Camembert, or vice versa?

The ethical and political aspects of this debate will be presented in detail in Part II (Ethics and Politics). Let us focus here on the problem concerning only its biological aspects: evolution, population genetics, and brain function.

The idea that the brain, unlike all other organs of the body, is not subject to natural selection, can be described as "selective dualism," or "liberal creationism" (Winegard,

Winegard and Anomaly, 2020). The latter term suggests that the brain is withdrawn from Darwinian laws to resort to a creationist vision motivated by liberal ethics. We fall back once again on this mixing of genres: ethics/scientific process. What do the data, the only source of scientific rigor, say?

IQ, a Cursed Statistic

This leads us to present the problem of the Intelligence Quotient (IQ) in a general way and, in doing so, to explain at length the problems of nature and nurture, and the often-misunderstood concept called heritability.

The "father of IQ" is the French psychologist Alfred Binet (1857–1911). The test he developed aimed to evaluate the intellectual capacities and mental age of each child compared to the rest of their class, with no idea of discrimination or exclusion. It was an educational tool, developed for the benefit of students. Subsequently, this test, imported to the United States and modified, became the Stanford-Binet Intelligence Scale, with clearly elitist aims.

The value of IQ as a measure of intelligence has long been controversial. The author of this book should not be considered a sharp "IQ-ologist." I have not been truly passionate about this debate, which is often very aggressive and tainted with ideology. I refrain from making a personal judgment on the question, the main reason being that it seems to me congenitally flawed by speculation and therefore doomed to go in circles. I content myself with reporting the main positions concerning it.

According to Franck Ramus, a researcher in psychology and research director at the CNRS, the main French research governmental agency (Bernichan, 2023), IQ remains the best measure we have of general intelligence: reasoning, comprehension, memory. But it does not cover certain cognitive traits like creativity, artistic talent, and social skills. For Plomin and von Stumm (2018), general intelligence can be defined very simply as the ability to learn, reason, and solve problems. It seems that IQ statistically has a strong predictive power on academic success and socio-economic success (Cochran and Harpending, 2009; Winegard et al., 2020).

It must be emphasized that the proponents of a hereditary component of IQ have never denied the impact of environment and education on cognitive faculties. On the contrary, the adherents of the "blank slate" ("environmentalist theory") denied any hereditary component of intelligence (Pinker, 2003), or at least considered it minimal. According to Pinker (2004), the "blank slate" hypothesis is no longer tenable in the face of biological advances. Winegard, Winegard and Anomaly (2020) are not so categorical and simply content themselves with presenting the "hereditarian theory" as more plausible than the environmentalist theory, in light of available data.

The Endless Quarrel of Nature vs. Nurture

The idea that the tendency to type 2 diabetes, cardiovascular diseases, or height have an innate component is hardly the subject of controversy (although...). What sets the stage on fire is the hypothesis that behavioral and cognitive traits (in this case, IQ) may have a genetic substrate, even a partial one. There is a major dividing line here between progressives and conservatives, between left and right. Throughout the 20th century, the "blank slate" theory, according to which the impact of the environment is determining, or even exclusive, dominated intellectual life, especially among social science specialists (Pinker, 2003, 2004). But even outside social sciences, some famous biologists (Stephen J. Gould, Richard Lewontin) considered that sociobiology (Wilson, 1978), a science whose object is the genetic command of behaviors, was intrinsically eugenicist and racist: "just-so-stories" (Alcock, 2017), that is, unconvincing, anecdotal explanations. The very possibility of not exploring the question of a possible genetic basis for behavior, whether it exists or not, was defended from a Marxist point of view by the famous biologists Lewontin, Rose and Kamin (1984). Rose (2009) defended the idea that research on IQ differences between individuals and between geographical populations should be discouraged: "it is nothing more than ideology dressed up as science." In fact, the argument can be turned like a glove against the author's statement. Ceci and Williams (2009) spoke out against this idea, which smells of censorship. The same goes for Richardson (2011), who philosophically recommended to "let the chips fall where they may."

The nature/nurture debate, when considering behavioral and cognitive aspects, is undoubtedly an incendiary subject.

What is the Current State of the Controversy?

> *No scientist maintains that the transmission of cognitive and behavioral traits is entirely explainable by genetics: all admit the importance of environment and education.*

To my knowledge, with the exception of a number of dishonest or ill-informed fanatics, who proliferate on social networks, no scientist currently maintains that behavior and intelligence are under exclusive genetic command (Pinker, 2004). As already stated, even racialist scientists who support the hypothesis that there are innate cognitive differences between human geographical populations ("hereditarian" theory (Winegard et al., 2020)); easily recognize that the environment plays an important role in the genesis of these differences (Arvey et al., 1994; Herrnstein and Murray, 1994; Rushton and Jensen, 2006; Sarich and Miele, 2004). The debate has shifted and now pits the adherents of the "blank slate" (the environment plays a predominant or exclusive role), who still represent a large and influential group, against those who believe that both the environment and genes have a significant impact on the genesis of cognitive and behavioral phenotypes (Berryessa and Cho, 2013; Pinker, 2003; Winegard et al., 2020). It is noteworthy that the American Society of Human Genetics states that "it is inaccurate to claim that genetics is a determining factor in human abilities," thus supporting the "blank slate" hypothesis (Foucart, 2025).

Is it Possible to Differentiate Between Nature and Nurture?

Not only is it possible, but it is indispensable. This is precisely the goal, for example, of behavioral genetics and the International Encyclopedia of the Social and Behavioral

Sciences (Knopik, 2015). Simply saying that environment and genes contribute to human nature is too often a strategy to sidestep the debate (see, for example, Jordan, 2008), which is counterproductive (Pinker, 2004). The respective contributions of genes and environment (heritability) must be evaluated, and the genes controlling the expression of the phenotype under study must be identified, as well as the different components of the environment concerned. Environmental factors can be, among others, the family, the extra-familial entourage, and chance (Pinker, 2004).

The evaluation of heritability is classically done by analyzing family pedigrees: identical twins raised separately or in isolation, fraternal twins, parents/children.

Identical twins are a godsend for geneticists and a curse for defenders of the "blank slate." Genetically, identical twins are comparable to clones since their genetic makeup is identical. Developmental mutations, after fertilization, constitute a notable exception. About 15% of identical twins show detectable genetic differences, which can be imputed to such mutations (Jonsson et al., 2021). Identical twins are an ideal tool for quantifying heritability, which is the part of the variability between individuals explainable by genetics (see glossary). Diverse experimental approaches, based on family pedigrees, allow the quantification of heritability: comparison between identical and fraternal twins, between identical twins, non-twin siblings and adopted children raised in the same environment, and identical twins raised separately. Animal experiments can also provide informative input. These experiments can, for example, use clonal species, which are genetically uniform, or lines of homogamic* mice (which are genetically very homogeneous) compared to other lines of homogamic mice, placed in diversified environments. However, this estimation of heritability by family pedigree analysis has been questioned by Lala and Feldman (2024), who insist on the importance of taking gene-culture coevolution into account. These authors note that parent/child cultural transmission can mimic genetic heritability.

The evaluation of the heritability of cognitive and behavioral traits must be done in light of gene-culture coevolution (Lala and Feldman, 2024).

Is Nature (Innate) Just a Matter of Genes and DNA Sequences?

Epigenetics refers to the phenomena of gene activation or inactivation in the absence of a change in the corresponding DNA sequence. It is increasingly thought that its impact on the genesis of phenotypic diversity in general, and on cognition and mental disorder phenotypes in particular, is considerable. The same goes for gene regulation, copy-number variation, and non-coding sequences (Tibayrenc, 2017a, c). Apart from the eventuality of developmental mutations, already mentioned (Jonsson et al., 2021), identical twins may not have evolved in exactly the same environment, which explains why they may not be perfectly identical, for example, in their fingerprints or size. This can be explained by epigenetic phenomena. The boundary between nature and nurture in this case is not clear, especially since some epigenetic modifications can be heritable.

Heritability of Phenotypes

Most phenotypes are the result of an interaction between genes and the environment. Phenylketonuria is a "Mendelian" disease (i.e., caused by a single gene), responsible for severe mental retardation if not detected in time. Its effects can be countered by a special diet, without the intake of the amino acid* phenylalanine. However, the idea that the effects of the gene (nature) can be totally corrected by the environment (nurture, in this case, the phenylalanine-free diet) is incorrect. Even if the special diet is correctly observed, children affected by this genetic disease have a low IQ, of about 0.80 to 0.90 (Pinker, 2004).

Many phenotypic traits are polygenic, meaning they are controlled by many genes, each having a limited individual effect. This is the case, for example, for height, which is highly heritable (Joshi et al., 2015). Twin studies show that the heritability of this trait is 0.89–0.93 (Silventoinen et al., 2003).

Heritability is a fiercely debated topic, even when limited to phenotypic traits that might be thought of as rather trivial. It was long considered, for example, that the heritability of skull morphology was low and that the environment played the main role in the genesis of this phenotype. This hypothesis was one of the battle horses of the American anthropologist Franz Boas (1912), a convinced environmentalist and a fierce opponent of eugenics. It is now considered questionable (Sparks and Jantz, 2002).

Heritability of Cognitive and Mental Traits

Thomas J. Bouchard (Dynesen, 2018) is an American geneticist and psychologist well known for his twin studies (the "MISTRA" program: Minnesota Study of Twins Reared Apart). Bouchard's studies (1998) emphasized the "Big Five" (main personality traits). These "Big Five" are extraversion, openness, agreeableness, conscientiousness, and neuroticism. These studies postulated that the heritability of these traits was between 0.4 and 0.6.

Eric Turkheimer, a professor of psychology at the University of Virginia, proposes three "general laws" on cognition (Dynesen, 2018): (i) all human traits exhibit some heritability; (ii) the impact of the environment on phenotypic traits is less than the impact of genetics; (iii) a significant part of the variation of most traits comes neither from genetics nor from the family environment. It is due to the cultural environment and chance (see also Pinker, 2004). It goes without saying that all these hypotheses, both those of Bouchard and those of Turkheimer, are far from consensual.

Estimates of the heritability of general intelligence range from 0.5 (Bernichan, 2023; Pesta et al., 2020; Plomin and von Stumm, 2018) to 0.8 (Anonymous, 2024b, Dobzhansky, 1973; Rushton and Jensen, 2006; Taguieff, 2018), or even 0.9 (Pinker, 2004). It is thought to increase with age ("the Wilson effect"; Dynesen, 2018).

The Specific Problem of IQ Differences Between Geographical Populations

Needless to say, the controversies over innate IQ and cognition differences between different geographical populations are fierce.

If we admit that IQ and cognition are the result of environment and genes, if the environment is made more homogeneous, as would be (should be) the case in democratic societies advocating egalitarianism, then the role of genes should increase (Pinker, 2003). However, this view is highly debatable. No society on earth today is truly egalitarian. People born into different social classes and different ethnic communities still face considerable disparities in living conditions and education.

The major debate focused on the innate or environmental origin of the "IQ gap" between African Americans and Americans of European descent. Racialist proponents of the genetic origin of these differences ("hereditarian" theory) include Arvey et al. (1994), Herrnstein and Murray (1994), Rushton and Jensen (2006), Sarich and Miele (2004), and Winegard et al. (2020). As already stated, it is necessary to emphasize that these "race realists" (Malik, 2012) recognize that the environment, education, and living conditions play an important role in the genesis of IQ and the differences in results between ethnic groups.

In the late 1990s, the American Psychological Association (APA) reconsidered the arguments presented in the best-seller "The Bell Curve" (Herrnstein and Murray, 1994) and was unable to refute them. The APA could only conclude that the genetic factors responsible for intelligence were still not clearly understood (Berryessa and Cho, 2013).

The controversy remains politically hot. I recall here the imprudent stance of ancient DNA specialist David Reich (2018a): "Is IQ, intelligence, or the number of years of study influenced by education? Of course. But does it also measure cognitive and behavioral characteristics? Almost certainly. And since all traits influenced by genetics differ from one population to another (...), genetic influences on behavior and cognition will also vary from one population to another." This statement triggered a real uproar (Mary, 2018a; Nelson, 2019).

Some scientists have gone so far as to make the rather insulting proposal of designing race-specific drugs that could improve the IQ of African Americans! (Richardson, 2011).

The numerical data, to my knowledge, are denied by no one (Grallet and Mahler, 2019). African Americans would have an average IQ around 0.85, while Americans of European origin would be at 1.00, and Americans of Asian descent, at 1.15 (Arvey et al., 1994). As already noted, Ashkenazi Jews have an average IQ higher than other populations of European origin, at 1.15. What there is no consensus on is the *interpretation* of these Figures. Environmentalists attribute the results to the environment, particularly education, economic level, and the stress caused by racism (Kaplan, 2015). Anyone who knows the United States (which is my case, I worked there for 5 years) cannot deny that the standard of living of the Black community still lags behind the rest of the population, despite the meritorious efforts of "affirmative action." This is not just a question of economic income. This gap affects education, the cultural environment, and access to care. However, for the opponents of environmentalism, the differences in IQ between ethnic communities retain an irreducible genetic component, even if they admit that the environment plays an important role. Indeed, in the genesis of brain function, no one denies that the share of the environment is crucial, contrary to the assertions of Ben-Ari, Clerget-Darpoux and Danchin (2023) and Clerget-Darpoux and Genin (2023), who claim that "hereditarians" stick to 100% genetic.

One hypothesis states that the intragroup heritability of IQ is strong, but that the heritability of *intergroup differences* (for example, between African Americans and Americans of European descent) obeys entirely different mechanisms. Even the "race realist" Arthur Jensen admits that the compared heritability of IQ between geographical populations could obey different mechanisms than the heritability within each geographical group (Dobzhansky, 1973). Such a mechanism is invoked by many (Harden, 2023). In the context of this debate, Winegard, Winegard and Anomaly (2020) cite the model proposed by Richard Lewontin to illustrate this hypothesis: two cultivation fields, A and B, one very fertile, the other very infertile, with environmental conditions being homogeneous within each of the fields, A and B. In A, the variability in crop size will be 100% heritable, and the same will apply in B. On the other hand, the differences in size between A and B will be entirely environmental. According to Winegard, Winegard and Anomaly (2020), such a picture is very far from the actual environmental differences observed between Blacks and Whites in the United States. These authors consider it more plausible to consider that a part of the IQ differences between the two communities is of genetic origin. To support these assertions, they cite the results of adoption studies by White parents: (1) of White children; (2) of mixed-race children, African American/European American; (3) of African American children (Minnesota Transracial Adoption Study; MTRAS). The IQ at 7 years old gives the following results: (1): 117.6; (2): 109.5; (3): 95.4; 10 years later: (1): 105.6; (2): 98.5; (3): 89.4. These results are more in favor of the hereditarian hypothesis than the environmentalist hypothesis.

The "cold winters theory" (Frost, 2019) was advanced to try to explain possible genetic differences in IQ between geographical populations. According to it, populations placed in very hostile environmental conditions would have developed increased cognitive capacities to meet these challenges. This theory has been ridiculed and considered "simplistic" by environmentalists. However, it is used with much less controversy when it comes to animals (Roth et al., 2012).

Some data complicate an already bushy debate that goes considerably in circles. The "Flynn effect" (Anonymous, 2021a) refers to a phenomenon of IQ increase from one generation to the next. It has been observed worldwide. Possibly related to this Flynn effect, the IQ of current African Americans is equal to that of Americans of European descent in 1947 (Ceci and Williams, 2009). Furthermore, the IQ of African Americans in the northern United States is higher than that of Americans of European descent in the South of the country, at least, this was the case in the 1950s (Dobzhansky, 1964).

The niche construction theory (Laland et al., 2010) says that organisms, when they modify their environment, modulate the natural selection acting on them. Because humans drastically act on their environment, they profoundly change the natural selection they undergo at the same time. *Culture* is the most obvious modification of the environment in the case of the human species. Depending on the type of cultural environment generated by this or that civilization (industrial, pastoral), the selection pressures generated will be different. This is the gene/culture coevolution theory (Winegard et al., 2020). We can recall here the theory as audacious as it is speculative proposed by Cochran and Harpending (2009) to explain the particularly high IQ of Ashkenazi Jews (see: "neurodegenerative diseases frequent among Ashkenazi Jews").

At the end of the presentation of this debate on cognition and behavior, it must be acknowledged that the results remain ambiguous and that it will be necessary to turn to other means of investigation to progress.

Moving Beyond this Endless Debate: Genomics and Brain Imaging

The advent of genomics has radically changed the dynamic of this hitherto sterile debate, which, as previously stated, could have "potentially explosive implications" (Richardson, 2011).

Right from the outset, it is important to emphasize that research into brain genes is still in its infancy (Cailloce, 2022). However, some informative results regarding possible differences between geographical populations have been gleaned.

The "star" of brain genes is the *FOXP2* gene, which is in fact a family of genes. FOXP2 is a transcription factor* linked to language and language disorders. Ayub et al. (2013) highlighted evidence of positive selection for the *FOXP2* genes in Europeans, evidence that would be absent in Africans and Asians. However, this hypothesis has been questioned by Atkinson et al. (2018). It appears that the *FOXP2* genes were affected by adaptive archaic Neanderthal introgression (Cochran and Harpending, 2009; Krause et al., 2007).

The microcephalin 1 (*MCPH1*) and *ASPM* genes, linked to brain development and growth, are other examples of brain genes with a differentiated geographical distribution and signs of positive selection (Laland et al., 2010). The links between these brain genes and general cognition are fiercely debated (Richardson, 2011).

The future developments of this research are absolutely impossible to predict. The possibility that substantial differences in brain function may be discovered between different geographical populations cannot be dismissed *a priori*. The considerable progress in medical imaging, genomics, and the study of ancient DNA allows us to anticipate new advances in this field.

Some recent data points in this direction of a geographical differentiation, mainly in the field of pathology. Benjamin et al. (2024) showed differences in gene expression of brain genes between geographical groups, in relation to different pathologies: strokes, Parkinson's disease, Alzheimer's disease. Similarly, the expression of genes related to schizophrenia appears to differ between European and non-European populations (Chen et al., 2024; Fu et al., 2024).

There have recently been definite advances in the research on the "intelligence genes" themselves.

Ji et al. (2025) compared two populations: Chinese (6,923 individuals) and Europeans (48,634 individuals) using a genome-wide association study (GWAS) analyzing volumetric traits of the amygdala, a brain structure linked to cognition, emotions, and certain neuropsychiatric disorders. They found 12 variants distinguishing the two populations (and two distinguishing the sexes). These variants "could explain the amygdala structural and functional differences between geographical populations."

Okbay et al. (2022) conducted a GWAS on 3 million individuals of European ancestry and identified 3,952 SNPs that allowed them to develop a polygenic risk score for educational attainment, predicting 12–16% of the variance of this trait. These authors note that GWAS developed on a European sample have a poor predictive power on an African population.

Sniekers et al. (2017) found 20 genes that are supposed to predict IQ, based on a study of 78,308 individuals. The fact that all these subjects were of European ancestry prohibits any comparison between geographical groups.

Liu and Liu (2023) identified 339 genetic variants associated with the volume of the hippocampus, a brain structure that plays a central role in cognition, memory, learning, and spatial orientation. Some of these variants are specific to given geographical groups.

Lee et al. (2018) analyzed 1.1 million individuals, all of European ancestry, with 1,271 SNPs involved in brain development and interneuronal connections. They developed "polygenic risk scores," which make it possible, given a certain combination of genetic markers, to predict the occurrence of a given phenotype. The phenotypes in question, in this case, were educational attainment and cognitive abilities. Their polygenic scores had good predictive power in the original sample, but less so in African Americans. This by no means implies that African Americans are intellectually inferior, but only that the genes involved in the development of their intellect are not exactly the same, if only in terms of frequencies, as those highlighted by Lee et al. (2018) in the sample of European ancestry.

Promoting Geographically Diverse Sampling Here and Elsewhere

This is therefore a research field that promises to be very fertile. Comparisons between geographical groups are a formidable research tool for finely exploring cerebral mechanisms. Denying ourselves this resource in the name of a misunderstood antiracism and egalitarism would have incalculable consequences for our knowledge of brain function, possibly to the benefit of other societies that do not bother with such misplaced scruples. Furthermore, beyond this research on brain function, the recommendation to diversify geographical sampling for all biomedical research is a major trend in current science (Murray, 2020; Wang et al., 2024). Indeed, until now, sampling based exclusively on individuals of European ancestry was predominant, which led to missing most of the genetic diversity of our species (see: "The 'genetic discovery bias'; too many 'whites'?"). The more or less clear calls for censorship

(see: Part II: "the temptation of censorship") do not seem to be heard, and that is fortunate. A constantly updated list can be found here of: (i) the numerous scientific publications that openly refer to the concept of race: (https://drive.google.com/file/d/1qWNbPeyBCwDL9CgzoOMrHguFm9MeX4f2/view); (ii) the even more numerous articles which, while replacing the term "race" with substitutes (ancestry, geographical populations, ethnic groups), constantly rely on standard sampling representative of the main geographical groups: Europeans, Africans, Asians, Native Americans, Oceanians:
(https://drive.google.com/file/d/1pffskqiM3WuPeD7gLIWYsSiXyTdwoeso/view).

Always Separate Morality and Science: Reciprocal Sanctuarization

Unfortunately, some people have a tendency to build hierarchies based on any difference. It is useful to recall here the view of the geneticist Theodosius Dobzhansky (1964, 1973), who insisted that equality is a moral, not a scientific, concept, and that it should not lead to considering all human beings and all populations as identical. This is also the opinion of Jordan (2008): "The term equality must not be confused with that of identity; on the contrary, it is because we are different that it is necessary to proclaim ourselves equal."

A recent report by the IBC (International Bioethics Committee; Gaydarska et al., 2024) can be again (see: "soft landings") cited here: " Present and future scientific discoveries are likely to reveal even greater fragmentation based on geography, race, and gender, allowing the genetic classification of not only health and disease but also personal traits, such as intelligence. As a result, the risk of discrimination may increase."

This is a major reason for resolutely basing antiracism on ethical and moral foundations, rather than on biology and genetics (Tibayrenc, 2017b, 2019; see Part II: ethics and politics). I will return to this at length in Part II.

Conclusion. Biological Race: The Return?

It would be a great exaggeration to say that the concept of biological race has regained full acceptance in the current scientific, and especially, political and ideological debate. However, despite the cautious denials of many authors who use them, it must be admitted that the concepts of "ancestry" or "geographical populations" closely resemble synonyms for the concept of geographical race (or subspecies) developed by physical anthropology (Coon, 1962; Vallois, 1976) and by population genetics (Ayala and Kiger, 1984; Dobzhansky, 1962a, 1973; Hartl, 1985; King and Stanfield, 1990; King et al., 2012; Sewall Wright, 1976, cited by Anonymous, 2024a). This is strikingly illustrated by the observation of Amuzu et al. (2025): "Genetic ancestry is a quantitative measure of inherited genetic variation. *and correlates with human migration patterns*" (my emphasis). We should recall that this concept of geographical race or subspecies is at the antipodes of the typological definition of race constantly put forward by scientific antiracism, namely, discrete entities (with sharp boundaries), each showing limited diversity (see "the typological concept of human races"). According to their definition, subspecies/geographical races can exhibit considerable genetic diversity, and may very well not have sharp boundaries, since they are potentially interfertile. The only clear boundaries that could be conferred upon them would be caused by geographical isolation. As soon as they become sympatric*, they give rise to intermediate populations, such as the two subspecies, *Lymenitis arthemis arthemis* and *Lymenitis arthemis astyanax* of the butterfly *Lymenitis arthemis* in North America (Anonymous, 2024c), the hybrids between the European wild boar (*Sus scrofa*) and the domestic pig (*Sus scrofa domesticus*), or the crosses between the wolf (*Canis lupus*) and the domestic dog (*Canis lupus familiaris*).

This is not to assert that the major human geographical groups *are* subspecies or geographical races. In the absolute, one can say neither that they are nor that they are not. It is a matter of definition. The concept of subspecies, or geographical race, is just a model. One is not forced to adopt it. Its general validity is debated by some authors (Lherminier and Solignac, 2005). This was used as an angle of attack by Livingstone (1962) to disqualify the concept of race in the human species. But ultimately, the fact remains that this concept of subspecies, or geographical race, has been fully validated by the international commission of zoological nomenclature (Lherminier and Solignac, 2005).

Furthermore, reasoning by *reductio ad absurdum*, a species is called "monotypic" in the case of a morphologically very uniform species. Under these conditions, it would be rather audacious to say that the human species is monotypic. The opposite of a monotypic species is a "polytypic" species, which, by definition, is composed of several geographical variants, or subspecies, or geographical races (Lherminier and Solignac, 2005). Q.E.D.?

According to Taguieff (2018), the consensus on the absence of biological race has significantly eroded since the beginning of the century. It should be recalled that this consensus has never been total. It has been accepted primarily in "Western" countries (North America and Western Europe), but, as already noted (see: "the consensus is dwindling"), this is not at all the case in Eastern Europe, Russia, and China (Anonymous, 2021a; Štrkalj, 2007).

Numerous authors in biomedical research have based their sampling on the concept of race, explicitly using the term, and continue to do so: (https://drive.google.com/file/d/1qWNbPeyBCwDL9CgzoOMrHguFm9MeX4f2/view), which shows that the consensus on the non-existence of biological race, put forward by many, is far from total. Finally, even more numerous authors: (https://drive.google.com/file/d/1pffskqiM3WuPeD7gLIWYsSiXyTdwoeso/view), while avoiding the use of the term "race," or qualifying it with cautious quotation marks, clearly base their sampling on the main geographical groups: Europeans, Africans, Asians, Native Americans, etc. This reveals an almost schizophrenic approach to biological research which, on the one hand, very often continues to officially refute the concept of biological race (not always), but on the other hand, bases its sampling on entities "that generally correspond to socially identified races" (Tishkoff and Kidd, 2004). It is indeed clear that what the terms "ancestry groups" and "geographical populations" cover is not truly distant from the concept of subspecies or geographical race.

As long as we only have to lament the discrepancy between the “official” motto (no biological race) and the more pragmatic attitude of biomedical researchers, we are dealing only with an academic debate with limited consequences, even if this inconsistency is damaging in terms of logic. Where the problems become more worrying is when this translates into conduct and decision-making in medicine (see Part II, Chapter 3: “The threat of scientific and medical obscurantism”).

Part II

Ethics and Politics

Introduction

Racial Theory, at the Heart of Racism: Biological Race Dictates Culture and Cognitive Abilities

> ***The less intelligent the white man, the stupider the black man seems to him. (French writer André Gide, Voyage au Congo [trip to Congo], 1927).***
>
> ***"The normal state of human beings is to be white, male, and heterosexual. All others do not belong completely to human nature" (Hull, 1986; cited by Kronfeldner, 2017).***

Disclaimer to the Reader: Part I (Biological Race) was based on my specific field of competence, namely medicine, particularly tropical medicine, evolutionary genetics, and taxonomy. In addressing the fields of ethics, morals, and politics, I am an ordinary citizen who expresses personal views, with a minimal philosophical background, I admit. However, these reflections have nonetheless been nourished by my long personal transcontinental journey (see introduction) as well as by my knowledge of contemporary history, a discipline that is a very dear hobby of mine and has been for a long time. Furthermore, the various developments set out below, which relate more to the humanities, will be constantly compared with what constitutes the most robust anchor point of this book, namely the description of human populations through the lens of population genetics, evolutionary genetics, and palaeogenetics. These disciplines have remained very close to a cautious description of raw data and are for this reason less subject to debate than other far more speculative approaches.

The term racism was used as early as 1892 by the anti-Semitic French journalist Gaston Méry (Chemin, 2022). The American definition of "racism" from the Cambridge Academy Content Dictionary emphasizes two essential dimensions of the term: hierarchization and discrimination: "the belief that some races are better than others, or the unfair treatment of someone because of his or her race." The definition given by the French Academy dictionary is: an "ensemble of doctrines according to which the varieties of the human species called races, principally distinguished from one another by their physical appearance, are endowed with unequal intellectual and moral faculties, directly linked to their genetic heritage." This is the exact statement of the "racial theory."

Racial theory, the basis of racism, which has been described as "racial essentialism" (Anonymous, 2024a) or the "hereditarian hypothesis" (Lasker et al., 2019), does indeed infer that biological race is characterized not only by physical or genetic traits, but also by intrinsic moral and cognitive properties, specific to each geographical population (Arvey et al., 1994; Rushton and Jensen, 2006). The naturalist Carl Linnaeus held such a view of races and deduced a hierarchy between them (Boqueho, 2023). This was a very widespread view at that time, a view that persisted until the 20th century. The famous paleontologist George Cuvier (1769–1832) and the naturalist Geoffroy-St Hilaire (1772–1844), both French, similarly inferred links between bodily differences and moral and intellectual capacities (Mary, 2018b). The same applied (Patou-Mathis, 2013) to the German physician and anthropologist Johann Friedrich Blumenbach (1752–1840). The "pope" of racial theory would be Count Arthur de Gobineau and his famous "Essay on the Inequality of the Human Races" (1853). The specificity of Gobineau's racial theory was to place the "Aryans" (Nordic Europeans) at the very top of the hierarchy, whom he saw as the founders of the most advanced and dominant civilizations. As already stated, this supposed superiority of the "Aryan race" was also defended by the French anthropologist Georges Vacher de Lapouge. Charles Darwin himself, although moderate (for his time) and an abolitionist, also believed in a hierarchy of races (Anonymous, 2022b).

In the contemporary era, the French anthropologist Henri Vallois, author of a famous book on human races (1976), attributed specific psychological characteristics ("their intelligence is flexible or sharp, or on the contrary, slow and lazy") to geographical populations (Hebert, 1977). The French hemotypologist Jacques Ruffié (1983) noted that in one school geography textbook from 1934, "race" was associated with intelligence. The German eugenicist Eugen Fischer (1874–1967) still supported the idea of an association between race and intelligence in 1943 (Ruffié, 1983).

The American anthropologist Earnest Hooton (1926) called the proponents of racial theory "ethnomaniacs." Without ruling out the possibility of cognitive differences between geographical populations, this anthropologist stressed that there was no scientific proof of such differences.

For the French anthropologist Claude Lévi-Strauss ("Race and History", 1952), the problem did not arise. For him, the only differences between ethnic groups were purely cultural. Despite all the respect I have for this immense scholar, I would say that he did not possess the necessary competence in biology to allow him to decide in this manner.

For many, merely evoking the eventuality of cognitive differences between geographical populations, even without inferring any hierarchies, is in itself a racist attitude. I have already discussed the case of the famous specialist of ancient DNA

David Reich, who was quite severely "reprimanded" for venturing to bring this eventuality to the table.

But this attitude of negation amounts to shut down the debate upstream, closing the question before it has been tested, and this, not within the framework of a scientific approach, but in the name of a moral and ethical approach, if not ideological. However, according to the guiding principle of this book, ideology should not have a place in science. We are faced with a *factual* question: innate cognitive differences or not between geographical populations. Two attitudes are possible: either one considers that the question is fraught with dangers and threats, and one explicitly decides to discourage such research. This is a perfectly acceptable attitude, and one cannot condemn it *a priori*. Such a recommendation was issued by Rose (2009) and Kaiser (2023). Or, on the contrary, one thinks that science must be free, that there is a very promising field of research here, that we now have the means to test this hypothesis (brain imaging, genomics, ancient DNA), and above all, that *differences* does not mean *hierarchy* (Tibayrenc, 2022), that many aspects of brain functioning are to be explored, apart from "intelligence" (altruism, empathy, personality, group strategies, musical sense...).

My feeling is that if we forbid ourselves such research, other countries, with different ethical rules, may not have so many scruples, and risk gaining a decisive advantage in this crucial field. In China, for example, an ambitious program to sequence the DNA of people with a high IQ (greater than 160) was launched in 2013 at the Beijing Genomics Institute (BGI), in Shenzhen, with the aim of selecting embryos supposedly carrying "intelligence genes" (Alexandre, 2013; Yong, 2013). The idea is to identify the genes for intelligence to help parents select embryos that carry them. BGI thus believes it can increase the intelligence quotient by 15 points in each generation (Project Code Name: Cognitive Genomics Projects; Grallet, 2018). The program is under the scientific direction of American physicist Stephen Hsu, from Michigan State University. Knowing that estimates of the heritability of general intelligence range from 0.5 (Bernichan, 2023; Pesta et al., 2020; Plomin and von Stumm, 2018) to 0.8 (Anonymous, 2024b; Dobzhansky, 1973; Rushton and Jensen, 2006; Taguieff, 2018), or even 0.9 (Pinker, 2004), this project is not necessarily utopian. Furthermore, the Chinese might be tempted to deepen their quest by enriching it with comparisons between geographical populations.

Lahn and Ebenstein (2009), for their part, enthusiastically invite us to "celebrate human genetic diversity" and, taking advantage of the ethnic richness of the United States, to encourage comparative research (not only in the cognitive field) between geographical populations. Behavioral geneticist Abdel Abdellaoui (Amsterdam University Medical Center; cited by Kaiser, 2023) considers these studies "inevitable." Ceci and Williams (2009) align with this idea of free research, but with the unusual recommendation of requiring a higher standard of quality for such a subject. In the "The international journal of science" section (2023), the journal *Nature* suggests the same: it is recommended that the highest standards of rigor be applied to studies taking "race, ethnicity and other socially constructed characteristics" as units of analysis. In other words, we would have two-speed research, a less demanding one for the genetics of the entire living kingdom except humans, and a more demanding one for humans! In my opinion, this is unacceptable. The living world is one; humans are an animal and a mammal like any other from a biological point of view, and must therefore be studied with exactly the same tools as the rest of the living world. On the other hand, the calls to formulate the results obtained by this field of research with all the desirable caution and nuances (Abdel Abdellaoui, cited by Kaiser, 2023; Chneiweiss, 2019; Richardson, 2011) can only be approved.

Chapter 1

Racisms?

Tocqueville's Paradox of Increasing Dissatisfaction: Racism Everywhere

The French philosopher Alexis de Tocqueville (1805–1859) is primarily known for his famous essay, "Democracy in America" (1835). Within the framework of his reflection on democracy, his "paradox of increasing dissatisfaction" dictates that the better a society functions, the more the feeling of what is wrong (e.g., persistent privileges) increases. According to political scientist Eric Kaufman (cited by Sastre, 2021), more and more conduct is being labeled as racism—at first glance, benign actions (for example, asking someone about their origin based on their physical appearance)—while truly racist actions (discrimination, assaults, insults) decrease. Abroad, I have often been questioned about my origin due to my physical appearance (Bolivia, Thailand) or because of my accent (United States). This has also happened in France to my Thai wife, especially since she habitually wears a very traditional Thai costume. I have always received these questions as simple, benevolent, and non-intrusive curiosity.

Regarding unequivocally racist behaviors, according to the Gallup Institute, in 1958, only 4% of Americans of European descent accepted mixed marriages. By 2013, this percentage had risen to 83% (Sastre, 2021).

It therefore seems clear that while authentically racist behaviors tend to decrease, the feeling of racism is becoming exacerbated.

Socio-Economico-Racial Stratification Worldwide

Notwithstanding these reassuring observations, without taking a large risk of error, one can surmise that, globally, there is a strong correlation between skin color and socio-economic level. In the United States, despite the meritorious efforts of "affirmative action" policy, the standard of living of the African American population remains significantly lower compared to other communities. Even in a country like Brazil, which prides itself on a racially very tolerant policy, I imagine that the palette of skin colors among dockworkers and laborers, on the one hand, and among politicians and businessmen, on the other, is not exactly the same. Furthermore, this concept of an ideally multiracial and mixed-race Latin America has been denounced as an illusion

DOI: 10.1201/9781003647393-7

and a strategy to continue to discriminate against populations of African and Native American descent with impunity (Rodriguez Mega, 2021).

These persistent socio-economic inequalities linked to ethnic origin certainly have infinitely complex, multi-factorial causes of an economic, sociological, and historical nature. However, for "race realists" (Herrnstein and Murray, 1994), these inequalities are primarily the result of genetic differences. This hypothesis is openly revisited by a world map indicating the supposed distribution of IQ by country, a map widely circulated on social media (Anonymous, 2023). It is drawn from the work of Lynn and Vanhanen (2002): "IQ and the Wealth of Nations". The main thesis of this book is that there is, not only a correlation but also, and above all, a cause-and-effect relationship between the gross national income per capita and the average IQ of the corresponding population.

In the current state of knowledge, it is impossible to absolutely exclude any biological component in these transnational IQ distributions. We fall back into the neverending and sterile debate about the origins of IQ (see Part I: recent evolution of cognitive traits). It is obvious that many environmental factors, particularly the quality of schooling, can interfere with such results.

Palettes of "Racisms"

"Acceptable" Racisms?

Nowadays, few people dare openly claim to be racist in the classic sense of the term (that people of color are intrinsically, biologically, globally inferior). The "cool" substitutes for racism, more or less acceptable, include:

Ethno-differentialism

According to this view, there are no inferior or superior human groups, but people from different cultures should not mix. This is the position defended by the French "Nouvelle Droite" (New Right), a right-wing school of thought (de Benoist, 1979). Proponents of this school of thought are characterized by a strong rejection of the virtues of miscegenation (Chemin, 2022), a rejection that Taguieff (2018) called "mixophobia." The work of Count Arthur de Gobineau (1853) was an extreme case of this mixophobia, the opposite of which is called "mixophilia" (Taguieff, 2022), which aims to encourage interracial crossings to achieve the real unity of the species. Hemotypologist Jacques Ruffié (1983), without going as far as advocating mandatory miscegenation, envisions a "white-coffee" mankind, with the advent of a relatively uniform population, tanned, with subtly slanted eyes.

Racialism

According to this school of thought, there is no hierarchy of races, but it is acceptable to describe human races and use them as units of study, for example, in biomedical research ("race realists": Malik, 2012). It is recalled here that anthropologist Ruth Benedict, a student of Frank Boas and a Figure of antiracism between the world wars, considered the study of races not to be racism (Taguieff, 2018). Racialism can be seen as positive if it highlights the evolutionary advantage of human genetic diversity (Klarsfeld and Vernier, 1998). For some, racialism is merely a disguise serving as an alibi for racism, and the two terms are equivalent.

As I have already mentioned, in the most recent biomedical literature, numerous publications openly refer to the concept of race in the biological sense of the term: (https://drive.google.com/file/d/1qWNbPeyBCwDL9CgzoOMrHguFm9MeX4f2/view). A recent, previously cited study, however, shows that the isolated use of the word "race" in articles published in the *American Journal of Human Genetics* (the organ of the American Society of Human Genetics) decreased from 26% to 4% between 1949 and 2018 (Byeon et al., 2021), in favor of more benign substitutes: "ancestry," "ethnicity," "race, ethnicity, ancestry" (REA), "Africans, Asians, Europeans." However, as already emphasized, on a conceptual level, these different terms are not truly far removed from the definition of the subspecies or geographical race, which is that of population genetics and physical anthropology. There is thus a semantic caution rather than a real paradigmatic shift. Moreover, this statistic is limited to a single journal, not necessarily representative of the entire biomedical literature.

Jordan (2021) agreed that the term "ethnic group" is readily used "to be politically correct."

Supremacism, With or Without Hate

This enters the zone of the "less and less acceptable," of out-and-out racism. Supremacists, who are vocal in the United States, defend the idea that the European-American community must continue to occupy the top position, as has long been the case. An odd convergence has been noted (Taguieff, 2018) between Neo-Nazi white North American separatists (George Lincoln Rockwell) and their black equivalents (Nation of Islam, Malcolm X), as if the two frenemies agreed to bury multiracial society and the "living together" ideology.

Racism by the Racialized: Woke Culture, Decolonialism, "Systemic" Racism

This is a very fashionable theme, a true leitmotif, not only in contemporary media but also in scientific literature. This school of thought is so powerful that it deserves a specific development.

Woke Culture

"Woke culture" (woke = awakened) intends to expose all the more or less concealed discriminations suffered by "minorities": women, LGBTQIA+ communities (for lesbian, gay, bisexual, transgender, "queer" and questioning, intersex, asexual, pansexual), and particularly, the "racialized," meaning non-white people. In short, everything that is not a straight white male. Non-whites are "BIPOCS" (Black, Indigenous, People of Color; Cheng et al., 2023).

A major, and possibly regrettable, trend of this woke culture is the *essentialization* and *racialization* of social relationships. Everyone is characterized above all by their skin color, and the entire structure of society stems from this.

The woke phenomenon has been masterfully analyzed by Taguieff in numerous writings (2018, 2020, 2022–2025).

This is not to deny that the different minorities mentioned above are subject to discrimination. Critics of woke culture do not attack certain realities brought to light by this "culture of awakening," but rather certain excesses, sometimes caricatural, of this culture. In this incessant victimhood, there is an attempt to seize power through the culpabilization of the dominant minority.

"Systemic Racism" and the Repentance it Engenders; "Critical Race Theory"

Some major scientific journals are endless in their acts of repentance for a racism that is supposedly omnipresent, consubstantial ("systemic") to "white" society in general, and to science in particular. Thus, for *Nature* (Anonymous, 2022c), science "is steeped in injustice and exploitation." In this article, the Afro-British author Daniel Akinbosede urges the reader to recognize the racist roots of science. He notes that out of 20,000 science PhD grants awarded by a certain British body, only 1.2% went to students of African descent. The editorial board of the *American Journal of Human Genetics* (American Society of Human Genetics Board of Directors, 2023), the official organ of the American Society of Human Genetics, apologizes for the past involvement of the society and some of its leaders in racist and eugenic enterprises. Else and Perkel (2022) call for increased recognition of "systemic racism" within society, including in science.

French anthropologist Michel Agier (Kodjo-Grandvaux, 2025) denounces the racist "infra-thought" of the elites, an inheritance, he believes, of European colonial history. Since white people are the descendants of those who practiced racism, they are supposed to have a specific task to accomplish in the antiracist struggle.

This concept of "structural" racism, deeply anchored in society, constitutes the essence of "Critical Race Theory" (CRT). According to this theory, racism is a "normal" category of society. The attitude of American voters towards Critical Race Theory shows a deep divide, as 94% of Donald Trump›s voters believe it is wrong to teach it in school, compared to 7% of Joe Biden›s voters (Le Bars, 2021). This cleavage will not have improved since Donald Trump's re-election.

Mervis (2022) openly accuses the referees deciding the allocation of funding from the American National Science Foundation of systemic racism, as white people are still overrepresented among the beneficiaries of these grants. The author does not specify how the referees could be informed of the candidates› skin color.

In the columns of *Science* (News at a glance, 2023), scientists from the American Society of Human Genetics (ASHG) express contrition for their association's past links with racism and eugenics up until the 1990s.

According to Popejoy (2021), racism "distorts science."

Such somewhat ostentatious "religious confessions" have been denounced by scientists Bikfalvi and Kuntz (2020). These authors encourage resistance to North American "cultural imperialism" and its "obsessive racial tribalism."

There is no doubt that science in the past, particularly anthropology, strongly endorsed colonialist and slave enterprises. On the other hand, it is highly debatable that current science continues to bathe in this "systemic racism." However, these attacks continue unabated. Consider this:

For Bindon (2020), who denounces a "molecularization of race," white geneticists are afflicted with "color blindness" and "structural racism," which are as inherent to them as the paleness of their skin.

According to Chen et al. (2021), systemic racism is an integral part of laws, policies, institutions, and practices. "Aversive racism" refers to denied racism. "Internalized racism" corresponds to the more or less conscious acceptance of the racial hierarchy by the dominated.

Cheng et al. (2023) claim that "Western" science is rooted in colonialism, white supremacy, and patriarchy. These authors, whose article is published in a well-known biology journal (*Trends in Ecology and Evolution*), warn against "microaggressions" and, to counter them, recommend semantic revisions in the vocabulary of the biological sciences. This can border on the ridiculous. For example, the terms "invasive" and "alien" should be banned from general ecological studies, as immigrants might feel targeted. Of course, the names of scientific authors who correspond to controversial historical Figures are deleted. The moth *Lymantria dispar*, which carried the vernacular English name "gypsy moth," is renamed "spongy moth" because of the texture of its eggs. It is not certain that these semantic maneuvers effectively reduce racism.

Duncan et al. (2024) believe that the study of genetics is influenced by Eurocentric, white, male, heterosexual, economically privileged power. According to these authors, science is intrinsically socio-political. This leads to a supposed exaggerated importance given to Mendelian genetics and the central dogma of molecular biology (which states that information flow follows the DNA-protein-phenotype pathway), which tends to give exclusivity to determinism (everything is in the genes) and essentialism (groups are homogeneous and intrinsically different). These criticisms are obviously caricatural. The authors recommend an antiracist genetics education. This last proposal aligns with the major theme of this book: the questionable mixing of science/morality. The teaching of genetics should focus on genetic science, not on antiracism, which is a moral, ethical, and political undertaking, not a scientific methodology (see the general conclusion of the book).

White Fragility; White Privilege

The concept of "white fragility" was introduced by Robin DiAngelo (2011), a North American sociologist. It refers to the denial, defensiveness, and guilt that white people produce when faced with conflicts of a racial essence. For DiAngelo, every white person is intrinsically racist, and if they deny it, that is proof that they are racist (Chavet, 2020).

"White privilege" requires little explanation: it refers to all the supposed or real advantages conferred by the simple fact of having light skin. Based on my personal experience, I would say that this concept is not unfounded. Whether in Bolivia or Thailand, two countries where I have stayed for many years, I have observed that the status of "gringo" (Bolivia) or "farang" (Thailand) did not only have disadvantages.

But this white privilege, constantly highlighted by "woke culture," tends to take on the appearance of a genuine religion, or even "antiracist racism." For the African American linguist John McWhorter (cited by Chavet, 2020), "Antiracism has replaced Christianity. White privilege replaces original sin as the stain that can never disappear." According to Françoise Vergès, daughter of the famous French lawyer Jacques Vergès, cited by Taguieff (2020), white privilege, which is present even among the poorest, must be deconstructed. Liliam Thuram, a famous footballer of Guadeloupean origin, vigorously defends the existence of this white privilege, contradicted in a lively debate by French philosopher Pascal Bruckner who, for his part, deplores the excessive culpabilization of "whites" (Strauch-Bonart, 2020).

Anti-White Racism

Badeau (2023) notes that sociologists and antiracist activists reject the concept of anti-white racism for the constantly put forward reason that racism presupposes discrimination, and that white people are never subject to any ostracism. This argument is also advanced by Graves and Goodman (2022). Badeau (2023) cites political scientist Daniel Sabbagh, according to whom racism has three dimensions: (i) ideological; (ii) affective; (iii) systemic. Therefore, for Sabbagh, anti-white racism exists because criteria (i) and (ii) are verified. In the same article, it is noted that LICRA (Ligue internationale contre le racisme et l'antisémitisme; International League Against Racism and Anti-Semitism) recognizes anti-white racism. According to Taguieff (2020), racialized minorities are slipping towards an anti-white racism while denying it—a true antiracist racism. When the rapper Nick Conrad calls to "hang the whites" (Taguieff, 2020), one is entitled not to find that entirely benevolent.

Decolonialism

According to sociologist Pierre Gaussens (cited by Bousenna, 2024), the "decolonial" ideology, denounced in this article as an "intellectual fraud," was initiated in North American universities by researchers of "Latino" descent. This ideology would take 1492—the arrival of Christopher Columbus in America—as its pivotal date. Coloniality is seen as a system of racially based dominance aiming to annihilate previous cultures. India also saw a rise in "postcolonial studies" in the 1980s. In these approaches, which resolutely turn their backs on the universalism of the Enlightenment, pre-colonial identities are idealized and mystified, and their flaws (patriarchy) are overlooked. As for the "West," it is demonized to the extreme.

Nobles et al. (2022) call for a "decolonization" of science and education.

According to Taguieff (2020), these "decolonial studies" do not constitute a structured field of study but proceed rather through slogans and clichés, promoting a rudimentary ideology. The strategy employed to give an appearance of seriousness is to mix genuine researchers with activists. Decolonialism, which is Afrocentric, considers Africa to be the true mother of civilization and denounces the white conspiracy of "falsification of history." Decolonialism considers France to remain a colonial state. Climate change is attributed to patriarchy, colonialism, racism, and capitalism. Historical Figures are vilified—not only Southern generals from the American Civil War but also personalities such as Ulysses Grant, George Washington,

Abraham Lincoln, Victor Schoelcher (who abolished slavery in France), Jules Ferry (the minister who promoted public education in France), Colbert (a minister of Louis the 14th), and General de Gaulle (Taguieff, 2020).

Is "Wokism" Effective Antiracism?

Do these incessant "woke" offensives serve the cause of antiracism? It is certainly the case that they have changed the essence of this noble struggle. Royer (2024) and Taguieff (2020) note that "historical" antiracist associations like the MRAP (Mouvement contre le Racisme et pour l'Amitié entre les Peuples; Movement against Racism and for Friendship between Peoples), SOS Racisme, and LICRA (Ligue internationale contre le racisme et l'antisémitisme; International League Against Racism and Anti-Semitism) now appear outdated and "old-fashioned", because they defend a universalist antiracism, in the spirit of the Enlightenment, and do not adhere to decolonialism, identitarianism, and the racialization of societal relations.

According to a pessimistic view, we are in fact facing "antiracism gone mad," according to the shocking title of a recent essay by Pierre-André Taguieff, analyzed by Roger-Pol Droit (2021). The new antiracists resolutely turn their backs on "classic," universalist, republican antiracism, stemming from the Enlightenment, which aimed only to combat discrimination without obsessing over skin color. On the contrary, the new antiracists resolutely divide the world into white and non-white ("racialized") and assign all individual destinies to the color of the epidermis.

A parallel can be drawn with the current divide between the two feminisms. The feminism that can be called "historical" unambiguously appealed to universalism and the Enlightenment. It denounced strict biological determinism. These are the famous "one is not born a woman, one becomes one" from French philosopher Simone de Beauvoir (1949), and the questioning of maternal love by another French philosopher, Elizabeth Badinter (1980). However, it avoided any overly victimizing approach (Badinter, 2003). Current neo-feminism presents the ostracization of the female gender as a branch of racism, a victim of white, Western, heterosexual patriarchy, within the broad framework of "intersectionality". This very fashionable term covers the Fight against all discrimination based on gender, race, social class (Le Bars, 2021)—in short, all the victims of the white patriarch. Multi-victims, such as black women, are given a special mention (Taguieff, 2020). This neo-feminism thus presents women and all gender minorities (homosexuals, lesbians, transgender people, etc.) as an oppressed category, on the same footing as the racialized, in an identitarian perspective. Muslim women who are asked to remove their veils in certain circumstances are considered to be oppressed by exacerbated secularism, viewed as the main weapon of state racism (Taguieff, 2020), whereas "historical" feminists present these veiled Muslim women as victims of male oppression. This identitarian neo-feminism goes so far as to relativize female sexual mutilation (excision) practiced in certain cultures, or even to defend it in the name of "cultural relativism" (Ahmadu et al., 2025).

The parallel between these two feminisms and the two antiracisms—the universalist and the identitarian—is therefore striking. These are major lines of cleavage within the current left, also contributing to blurring the distinction between left and right.

Chapter 2

Semantic Puzzle

What Should Replace the "Word Too Many": Race?

Current media outlets often feature columns intended to settle highly topical debates. These are, for example, the two French newspapers *Le Monde*'s "les décodeurs" (the decoders), and *Libération*'s "check news," or Google News's "fast-checking." These new oracles deliver verdicts in the form of standard answers: "false," "mostly false," "partly true." Faced with the question: "Has science proven the non-existence of human races?" I would cautiously suggest: "it is more complicated," an answer often given by our modern oracles. Indeed, it is more complicated. Biology can undoubtedly reject the typological and phylogenetic concepts of race, which do not correctly describe the biological diversity of our species. However, as I have emphasized, these concepts have never been current, neither in physical anthropology nor in population genetics. Furthermore, obviously, the social science concept of race as a pure sociological concept, the target of all discriminations, the alibi for colonialism and slavery, corresponds to nothing from a biological point of view. On the other hand, the concept of race as a subspecies or geographical race, the one that was current in physical anthropology and population genetics, can be debated, like any taxonomic entity. However, as a model, it has a rather satisfactory predictive power.

The debate actually boils down to a matter of definitions and semantics. For undoubtedly, the "word too many" (Taguieff, 2018) has a heavy "rap sheet," and severely shocks many interlocutors. The proposal to replace it with more acceptable terms for moral reasons is therefore entirely valid (see Part I: "The concept of human race must be rejected because it is dangerous"). There is thus a need for semantic innovation, especially since the debate can be accompanied by a good dose of intimidation against those who persist in using unwelcome vocabulary (see Chapter 3: "The temptation of censorship?").

But what satisfactory term can replace "race"? Many less controversial terms circulate, particularly in biomedical literature, but none are truly satisfactory.

DOI: 10.1201/9781003647393-8

To refocus the debate, I recall the definition of human races as subspecies or geographical races given by physical anthropology and population genetics: geographical subdivisions of the human species exhibiting certain phenotypic traits and certain gene frequencies that allow them to be distinguished from one another. The substitute term should have the same predictive power and be understood by the greatest number of people, not just a restricted circle of specialists.

Ethnic Group

This is the first "sweetener" that became common. It was reportedly advocated by Ashley Montagu during the post-war UNESCO meetings. It is still widely used. However, it is especially unsatisfactory for two reasons.

Firstly, one of its main proponents is the controversial French anthropologist Georges Vacher de Lapouge (1854–1936), a key propagandist of the "Aryan race," or Nordic race, theory, which stated that Northern Europeans, tall, blonde with blue eyes, and dolichocephalic (an oblong skull viewed from above) represent the most perfected humanity. Vacher de Lapouge was considered by the Nazis as one of their spiritual fathers.

Secondly, this author gave the term "ethnic group" its *correct* meaning, which many tend to forget. According to him, the ethnic group, unlike race, concerned culture, not biology. For Vacher de Lapouge, Germans and French were two peoples of the same race (biologically identical or similar) but corresponding to two different ethnic groups.

So, ethnic group: a controversial lineage and a misinterpretation. It could really be better.

Ancestry Group

This term designates individuals who share a common ancestry. It is often considered in the English-language biomedical literature as a very satisfactory substitute for the "word too many." In fact, if you think about it, it is extremely vague: my children are an ancestry group. Queen Victoria's descendants are an ancestry group. Genghis Khan's numerous offspring are an ancestry group (he is considered one of the humans with the greatest "evolutionary success" or "fitness"). Therefore, "ancestry" or "ancestry group" is only meaningful if one sets aside its strict definition and voluntarily limits oneself... to the well-identified groups that share certain phenotypic traits and certain gene frequencies.

The fragility of this euphemism is illustrated by the candid observation of Mauro et al. (2022), who note "the striking resemblance of continental ancestry categories to racial categories." The American sociologist Rogers Brubake (cited by Taguieff, 2018) also notes that the genomic differences associated with geographic populations are explicitly linked to popular concepts of race. One can also recall the observation of Amuzu et al. (2025): "Genetic ancestry is a quantitative measure of inherited genetic variation, *and correlates with human migration patterns*" (my emphasis).

Race/Ethnicity/Ancestry (REA)

This composite term, which has the appearance of a Solomonic judgment, suggests that the three terms composing it are more or less equivalent. This trinomial is increasingly used in English-language biomedical publications. See for example Ali-Khan et al., 2011; Arpone et al., 2024; Popejoy et al., 2018. Apart from the advantage of not having to use the "word too many" in isolation, the contribution of this trinomial, combining the problems of the meaning of "ethnic" and "ancestry," is not clear.

Geographic Population

This is yet another very popular substitute for the "word too many." It has the same imprecision as "ancestry group," but worse. Indeed, the latter still referred to a certain biological reality, whereas "geographic population" can be applied in all sorts of ways: New-Yorkers are a geographic population. The inhabitants of Canada are a geographic population. This term, therefore, only makes sense if it refers to subdivisions of the human species such as Africans, Europeans, and Asians.

Conclusion on the Semantic Puzzle

The various terms used to sidestep the "word too many" are rife with imprecision and misinterpretation. These drawbacks are only overcome if one implicitly refers to the notion of major geographical divisions of our species—in other words, more or less, to the concept of subspecies or geographical race. This is, in fact, what happens in the publications where these terms are used. And I confess: I am the first to do so in my publications. For why shock the reader? And why risk the immediate rejection of an article for reasons of semantic provocation? This self-censorship even operates within my intellect. In my reflections, I do not think "black race" or "white race." The "word too many" is decidedly very unwelcome. I have opted for "ancestry," adding "African," "Asian," "European," etc. If greater precision is required, within "African ancestry," I add, for example: "Fulani ethnic group". Perfection is not of this world, in semantics as elsewhere. The fact remains that, if I believe the publications concerned with these terms, the descriptions of what "ancestry" or "ancestral group" are truly very difficult to distinguish from the definition of subspecies or geographical race.

Chapter 3

The Threat of Scientific and Medical Obscurantism

The debate on the existence or non-existence of human races has reached an extreme virulence. However, when it is limited to a purely academic joust or semantic hairsplitting, it is only a partial evil. It is quite different when this debate comes to obstruct the scientific process, and worse, the reliability of medicine. Yet, this danger has grown in recent years, becoming a veritable "cancellation" of research and of the art of medicine.

It is legitimate to be greatly concerned when immensely prestigious North American medical societies, the American Medical Association (AMA) (Press Release, 2020) and the American Academy of Pediatrics (AAP) (News at a glance, 2022; Wright et al., 2022), solemnly call on their members and the medical profession in general no longer to take into account the ethnic origin of their patients in any medical process whatsoever. This position is shared by Graves (cited by Gaskin, 2025).

Why the Positions of the AMA and the AAP Are Medically Debatable

As explained in Part I (see Chapter 2: Geographical populations and medicine), many, if not most, diseases exhibit very different frequencies among geographic populations, and these differences have a genetic basis, even though the environment plays an important role in these differences. I will cite two extreme examples here, which are especially illustrative in the perspective of the AMA and AAP positions:

Sickle cell anemia (drepanocytosis) has a high incidence in African Americans, while it is virtually absent in Americans of European descent. Indeed, the genetic variant associated with sickle cell anemia has a 1500 times higher frequency in Africans than in Europeans (The UK Biobank Whole-Genome Sequencing Consortium, 2025).

For Cystic Fibrosis, the reverse is true: it is much more common in Europeans than in Africans.

DOI: 10.1201/9781003647393-9

This does not make sickle cell anemia a "Black disease" and cystic fibrosis a "White disease," according to the polemical argument of Graves and Goodman (2022) and Lala and Feldman (2024). No one has ever said that 100% of African American children are afflicted with sickle cell anemia, any more than 100% of White children suffer from cystic fibrosis. We are talking about *frequencies* here: the *probability* of diagnosing sickle cell disease in an African American child is much higher than making the same diagnosis in a White child. Discouraging a pediatrician from bearing this in mind is aberrant. The picture is reversed for cystic fibrosis.

In France, the Haute autorité de santé (High Authority for Health) recommended the abandonment of ethnic targeting (considered "stigmatizing") for newborn screening for sickle cell anemia (Brafman, 2024): it is now extended to everyone, even though the probabilities of identifying this genetic trait in children of European ancestry (still the majority in France) are minuscule.

The AAP's position is energetically supported by Wright et al. (2022), who elaborate in these terms: "The AAP will rely on the 'Words Matter' document to ensure that all authors, editors, presenters, media spokespersons, and other content contributors recognize race as a social construct only and refrain from any use or reference to it as a biological unit of study ('proxy'). All professional organizations and medical specialty societies should advocate for the elimination of race-based medicine in any form." The authors deny possible differences in lung capacity, kidney function, and the incidence of atherosclerotic heart disease, even though these differences are strongly supported (see below). They do not mention the particularly telling cases of sickle cell anemia and cystic fibrosis.

Questionable Positions of Other Medical Societies

"The current state of evidence from the American Thoracic Society and the European Respiratory Society has led to an official statement that race and ethnicity should no longer be considered in pulmonary function testing" (Press Releases, 2023). However, numerous data contradict this statement. All things being equal (environmental parameters), there are obvious differences in heritable pulmonary capacity among geographical populations (Johnston et al., 1987). Neglecting this parameter would lead to misleading conclusions about the pathological or non-pathological lung capacity of many individuals. An extreme example concerns populations that have adapted to high altitude. In Bolivia, the Aymara Native Americans, inhabitants of the Altiplano, have an increased lung capacity, which is anatomically visible: they have a "barrel chest." This trait was scientifically explored by the Instituto Boliviano de Biologia de Altura (IBBA; Bolivian Institute of High-Altitude Biology; https://ibba.umsa.bo/), a Franco-Bolivian institute where I worked for 7 years. It would therefore be completely misleading to consider that there are no differences in lung capacity between the Aymaras and other geographical populations. Aymaras are not the only geographical population where lung capacity is affected by genetic ancestry. Lung function exhibits differences according to ethnic groups and is affected by the degree of miscegenation, which impacts the incidence of asthma (Mersha and Abebe, 2015; Kumar et al., 2010).

Regarding kidney function, the position of the National Kidney Foundation, the American Society of Nephrology, and the relevant European professional organizations is presented by Delgado et al. (2021): it is simply proposed not to consider ethnic origin in the analysis of kidney function. However, Stevens et al. (2011) compared the glomerular filtration rate among five ethnic groups: Black, Asian, Native American, Hispanic, White and "others." They found significant differences among these groups: lower than White and "others" in Black and Asian people, higher in Native Americans and Hispanics. This is confirmed by the fact that neglecting ethnic origin can lead to a flawed estimation of the glomerular filtration rate in African Americans (Hsu et al., 2021). Moreover, kidney diseases in general (glomerulosclerosis, arterial hypertension) are more frequent in African Americans than in Americans of European descent. This has been attributed to variants of the Apolipoprotein L1 (*APOL*1) gene, which would play a role in protection against sleeping sickness (Esoh et al., 2025; Karlsson et al., 2014; Lachance and Tishkoff, 2013). It is therefore highly questionable to consider ethnic origin as having no impact on kidney function and kidney diseases.

As a physician, I believe that the strategy of these medical associations is quite debatable. One can suspect that they are driven by the concern to promote antiracism, which is certainly very honorable, but this should not be done at the expense of the rigor of medical science.

Scientific and Medical Obscurantism in Scientific Literature

We can recall here the undeniably false assertion made by Richard Lewontin (anonymous 2002b) that skin cancers only appear after reproductive age and therefore cannot be an agent of natural selection. However, melanoma is the most frequent cancer in young adults between 25 and 50 years old. This false hypothesis led him to assert that natural selection, through climatic adaptation (sunlight level), could not explain the differences in skin color between geographical groups. He explained these differences by sexual selection. However, the hypothesis of climatic adaptation of skin color is very generally accepted (Takezawa, 2023). This extreme attitude of Richard Lewontin is explained by his reluctance to admit any biological determinism whatsoever.

I have already explained in Part I the fallacious strategy used by many authors who disparage genetic studies, particularly genome-wide association studies, on the pretext that they neglect or minimize the impact of the environment on the genesis of phenotypes (see: "Sophist dichotomy between nature and nurture, between genetics and environment").

Omiye et al. (2023) have pertinently warned about the danger of race-based medicine being propagated by large language models (LLMs). One can only share the authors' concern about the potential misinterpretations by AI software dealing with innate differences between different geographic populations. However, several of the examples cited by these authors seem questionable. This reduces the potential impact of their otherwise commendable message. The most blatant example is the case of lung capacity, already discussed above (see: questionable positions of other medical societies).

The assertion that medical studies of kidney function based on race have been scientifically refuted is also debatable, as stated in the previous paragraph (analysis of the work by Delgado et al., 2021).

Omiye et al. (2023) also assert that there are no differences in pain sensitivity between ethnic groups. This has been disputed by Kim et al. (2017).

Even the trivial case of skin thickness (which the authors claim shows no difference between African Americans and Europeans) is debatable. In a recent meta-analysis study, such differences, although described as "minor" in 6, were observed (Table 4 in Lintzeri et al., 2022).

One can suspect that Omiye et al. (2023) were motivated by moral and ethical antiracist concerns when writing their article. This is entirely respectable. Systemic racism must absolutely be fought. However, this should not be done at the expense of scientific rigor. Contrary to biomedical research, where race/ethnicity/ancestry are very often considered as relevant units of analysis (see numerous references in: (https://drive.google.com/file/d/1qWNbPeyBCwDL9CgzoOMrHguFm9MeX4f2/view). (https://drive.google.com/file/d/1pffskqiM3WuPeD7gLIWYsSiXyTdwoeso/view)), there is a tendency in the medical associations cited above to recommend the opposite. In my opinion, such recommendations follow more moral and political motivations than medical ones, as seems to be the case with the article by Omiye et al. (2023). These approaches, although morally commendable, cannot dismiss with a stroke of a pen the numerous studies that contradict them (see numerous references in Tibayrenc, 2017c and in Part I).

Roberts (2021), in the columns of the prestigious British medical journal *The Lancet*, directly attacks all "race correction" (correction of medical quantitative indices based on ethnic origin), not only for kidney function, but also for respiratory, cardiological constants, the incidence of arterial hypertension, and breast cancer. In light of what has been explained above, and in Chapter 2 of Part I: "Geographic populations and medicine," this proposal appears as a flagrant example of medical obscurantism.

Anti-Obscurantists

Certain authors do not adhere to this approach, which is tending to become dominant in the world of medical associations. A particularly remarkable contribution is that of Oni-Orisan et al. (2021). Besides their qualities as eminent scientists, the five authors have the particularity of all being of African descent. They can therefore hardly be suspected of "systemic racism." They do not hesitate to break a major taboo by asserting that race (in English in the text and without quotation marks) has both a *genetic* (emphasis added by me) and a sociological component ("race has both a genetic and a social component"). Oni-Orisan et al. (2021) defend the idea that "race and ethnicity" categories must be considered in medicine and science precisely to promote better health equity for the benefit of the African American community. Indeed, the genetic variability within populations of African origin, including African Americans, is much greater than that observed in Europeans. Basing research on genetic data coming only from European sampling therefore leads to an ignorance of many genetic variants of major medical importance specific to African populations (see "the genetic discovery bias," Part I).

The Temptation of Censorship

For the past twenty years, certain messages have carried thinly veiled threats, or at least pressures, against those who would continue not to explicitly reject the concept of race. We will see that this temptation of censorship does not only concern the problem of race but extends to studies of gender, cognition, and behavior. This trend has been denounced by Coyne and Mareva (2023). This stance is all the more remarkable since the first author, Jerry A. Coyne, is a talented evolutionist, a student of the leader of scientific antiracism, Richard Lewontin. Coyne is known for his stances against apartheid in South Africa. He is little suspected of racist thought.

In the United States, the National Academies of Sciences, Engineering, and Medicine (2024) put pressure on what enables research—namely, the sinews of war: funding: "Journal publishers and editors, research funders, and sponsors should require researchers to provide a scientific rationale for their use of race and ethnicity, describe data provenance, and acknowledge limitations of their use." This organization advises against the use of the terms "race" and "ethnicity" in the titles and conclusions of scientific articles.

Figure 7 in Part I shows the results of a study describing the genetic variability of different geographical populations according to the latest standards of genomics (Bick et al., 2024). The journal concerned, *Nature*, received numerous requests for the retraction of this article, for the simple reason that the Figure clearly depicted the main human geographical populations (Asians, Blacks or African Americans, Middle Easterners or North Africans, Hawaiians or other Pacific Islanders, Whites) The authors refused to retract their article.

In her book "Superior: The Return of Race" (2019), science journalist Angela Saini states that the simple idea that races correspond to real genetic differences fuels white supremacism.

According to Berryessa and Cho (2013), we lack an ethical framework "to know whether there are some research questions that should not be asked at all." One could not be clearer: certain research topics could be purely and simply banned.

According to Brown et al. (2025), "Some scientists attempt to shield potentially injurious work from criticism by appealing to the 'value-free ideal' of science, which holds, incorrectly, that the ethical, political, and social implications of research can be separated from science. But academics have an obligation to report their research findings accurately and to consider how their research might be misused or misreported." The reader will see later (see the general conclusion) that I do not adhere to the proposition of Brown et al. (2025). Of course, scientists must present their results with all the necessary prudence and nuance but being overly concerned about all risks of misuse amounts to striking with suspicion, or even censoring, any research on "hot" topics.

Carlson et al. (2022) warn against the appropriation of genetic research by racists. They recommend that research teams be led by people trained accordingly against such instrumentalization of science. This proposal seems a bit like a "political commissar."

According to Cooper et al. (2003), scientists cannot evade the message carried by the concept of race. If genomics were to endorse this concept, it would be equivalent in the eyes of these authors to accepting the principle of a hierarchy between races. They consider the mere act of speaking about race to be a manifestation of racism. I recall here that Ruth Benedict, a student of Franz Boas (both leaders of North American scientific antiracism between the two wars), recommended distinguishing between the study of races and racism (Taguieff, 2018).

In the editorial of a special issue dedicated to the problem of biological race (2000), *Nature Genetics* issues a warning, addressed to both authors and reviewers, a warning that has the advantage of clarity: "From now on, *Nature Genetics* will therefore require that authors explain why they make use of particular ethnic groups or populations, and how classification was achieved. We will ask reviewers to consider these parameters when judging the merits of a manuscript—we hope that this will raise awareness and inspire more rigorous design of genetic and epidemiological studies."

An editorial in *Human Nature Behavior* (2022a), after a cautious remark insisting that academic freedom is fundamental but not without limits, is even more explicit: "We recommend avoiding the use of descriptors that refer to attributes such as race, ethnicity, national or social origin, sex, gender identity, sexual orientation, religion, political or other beliefs, age, disease, (dis)ability or other group descriptors unless they are relevant." The editorial clearly announces that an article may be retracted or refused if it contravenes these recommendations.

Xavier Jeunemaître, head of the genetics department at the Georges Pompidou European Hospital in Paris (cited by Rosier, 2018), explains that "intelligence is such a stigmatizing trait, and whose Intelligence Quotient (IQ) only captures such a limited part, that it would be better to refrain from such research." He was referring to the research of the highly controversial Robert Plomin.

Kaiser (2023) expresses a similar concern. According to him, due to scientific uncertainties, scientific journals and their sponsors should avoid comparative studies for "sensitive phenotypes."

Mauro et al. (2022) envision pressures against the reluctant in the form of article rejections or lack of funding.

Molteni (2022) pertinently recalls that the perpetrator of the Buffalo shooting in 2022 was an avid reader of scientific publications relating to differences between geographical populations. According to this author, the simple act of questioning the possible links between gene and behavior is "playing with dynamite."

Omiye et al. (2023) are alarmed that artificial intelligence software might propagate race-biased medicine. They recommend "adjusting this software" to eradicate these inappropriate racial terms (see part I: "The Highs and Lows of Artificial Intelligence").

For certain physiological functions and certain pathologies, there are statistical differences between the different human geographical populations. For kidney function, we cite the glomerular filtration rate; for lung function, spirometric tests. A given value of these different functions will be considered normal in one population and pathological in another (Tibayrenc, 2017c). Roberts (2021) asserts that these approaches are based on erroneous concepts of "racial" medicine dating back more than 50 years and proposes to simply eliminate the corrections for these different tests and to treat all geographical populations in the same way. In my opinion, this is a flagrant case of medical obscurantism (see Part II, Chapter 3).

Rose (2009) considers scientific research on human races (and on IQ) to be merely "ideology dressed up as science." According to him, scientific methodology must obey two criteria: (i) the question must be well posed; (ii) it must be possible to answer it with the available technical and theoretical tools. This is not the case, according to Rose, for research on human races (and on IQ, and on comparisons between genders). This research must therefore be discouraged, that is, censored.

As already cited, Wright et al. (2022) are explicit: "The AAP (American Academy of Pediatrics) will rely on the 'Words Matter' document to ensure that all authors, editors, presenters, media spokespersons, and other content contributors recognize race as a social construct only and refrain from any use or reference to it as a biological unit of study (proxy). All professional organizations and medical specialty societies should advocate for the elimination of race-based medicine in any form."

I am sharing a personal anecdote here. In the 1990s, I served as a project referee (for funding requests) for the National Institutes of Health (NIH), which, aside from their research activities, are a major funder of biomedical research. One of the funding requests ("grant applications") concerned the comparison of malaria susceptibility between Pygmies and Bantus. One of the evaluators present aggressively attacked the project, stating that the entity "Pygmy" had no biological reality and, therefore, that the scientific program, apart from carrying a dubious ideology, was fundamentally flawed. The message was received loud and clear by all present evaluators, who theatrically began tossing their copies of the proposal into the trash can. Alas, nowadays, this research topic has become utterly standard, Pygmies have recovered their status as a well-identified people with clear genetic specificities, and studies comparing them biologically to their Bantu neighbors are numerous (see, for example: Donnelly et al., 2016; Korzeniewski et al., 2021). The NIH censorship simply caused this field of research to be delayed by many years, and the author of the discarded project had to seek other funding sources.

Coyne and Maroja (2023) outlined six areas where an ideological view leads to a false understanding of the issues concerned:

1. In humans, sex is not a discrete, binary distribution of males and females, but a spectrum.
2. All behavioral and psychological differences between men and women are due to socialization.
3. Evolutionary psychology, the study of the evolutionary roots of human behavior, is a specious discipline founded on false assumptions.
4. We should avoid studying genetic differences in behavior among individuals.
5. "Race and ethnicity are social constructs, without scientific or biological meaning."

6. Indigenous "ways of knowing" are equivalent to modern science and must be respected and taught as such.

This inventory goes far beyond the topic of the present book. However, in the authors' eyes, they are all struck by the same fate: the subjugation of the scientific process by ideology, which is a major theme of the present book. According to Coyne and Maroja (2023), authors who might risk questioning the statements listed above are ostracized in two ways: either indirectly, through deterrence, or directly through direct sanctions (rejection of funding requests, rejection of articles, professional sanctions that can lead to dismissal).

Lee (2022), a behavioral geneticist at the University of Minnesota (United States), reports a supposed tendency of the prestigious journal *Nature Human Behavior* to reject manuscripts that do not follow a certain ideological orthodoxy. This author reports that the National Institutes of Health seems increasingly to limit access to the "Database of Genotypes and Phenotypes" (dbGaP), even for studies that do not include race or sex comparisons. Any research presumed to be "discriminatory" seems banned. According to Lee (2022), the NIH appears to be suppressing any study focusing on the associations between genetics and intelligence.

Faced with these attempts at censorship, we can fear a "Streisand effect", the phenomenon whereby information becomes more and more visible as attempts are made to censor it.

One cannot help but be struck by the symmetry with the wave of repression that has fallen especially brutally on American science since the last presidential elections in November 2024. Attempts to surveil foreign teams (Europe, United Kingdom, Australia, Canada) receiving American funding have been noted (Mallapaty, 2025). Some have received questionnaires asking if their research was associated with topics such as gender ideology, climate change, diversity, equity, and inclusion (DEI).

One ideology chases out the other, with potentially very negative effects on academic freedom in both cases. A middle line, maintaining equal distance from both ideologies (Taguieff, 2025), appears as necessary as it is difficult to hold. Robert (2025) and Tournier and Robert (2025) note that among the scientists speaking out within the "stand up for science" movement against the new policy of the American government, none spoke out to protest against the censorship, semantic directives, and sanctions taken against several scientists during the previous administration. Moreover, these sanctions were not limited to the United States, as they provoked the suicide of Richard Bilkszto, a college director at the Toronto Academy, Canada, accused of white supremacism (Tournier and Robert, 2025). This sad affair went largely unnoticed.

Scientific rigor risks being profoundly affected by this ideological grip. The best strategy against this peril is certainly to try to extirpate all ideology from the scientific process (see general conclusion).

This pressure from veiled or open censorship is not benign and certainly has an impact on the research field of human biological diversity. However, it is worth noting that despite the nevertheless clear warnings from *Nature* (2000) and *Nature Human Behavior* (2022a), the flood of biomedical publications using the major human geographical populations as units of analysis shows no sign of drying up, and even, that many of them do not bother with semantic circumlocutions, and clearly use the term "races" to refer to these geographical populations. (see links on: (https://drive.google.com/file/d/1qWNbPeyBCwDL9CgzoOMrHguFm9MeX4f2/view). (https://drive.google.com/file/d/1pffskqiM3WuPeD7gLIWYsSiXyTdwoeso/view)

Freedom of Expression?

In the face of this temptation toward censorship, some authors defend a certain freedom of expression, trying to preserve a fragile balance. I have already cited the editorial in *Human Nature Behavior* (2022a), and its rather hypocritical declaration that "academic freedom is fundamental but not without limits," only to delineate the limits a little further on.

Berent (2022) speaks out against the temptation of censorship. According to this author, limiting science will never alleviate social ills; on the contrary. Science alone is capable of correcting the injustices of the "genetic lottery."

Ceci and Williams (2009) advocate for free research, but with the unusual recommendation of requiring a higher standard of quality assurance when the research concerns human geographical diversity. No censorship, but a little bit anyway.

In the column "The international journal of science" (2023), the journal *Nature* makes a similar recommendation: that the highest standards of rigor be applied to studies using "race, ethnicity, and other socially constructed characteristics" as units of analysis.

An editorial published in *Nature Reviews Genetics* (2022b) defends a perilous balance with more conviction: "Scientists must indeed be more aware of how their findings can be distorted, but they must be free to pursue their research without fear of repression or censorship. Science, and the study of human genetics in particular, are not allies of fascist and racist views."

The "Saint-Simonian utopia" (from the French philosopher Claude-Henri de Rouvroy de Saint-Simon (1760–1825)) is a school of thought stipulating that science can rid human societies of all their imperfections. According to this school of thought, since the engine of social progress resides in industry, technology, and science, society should be governed by scholars, industrialists, and engineers. This is an extremely scientistic and technocratic view of society. The French journalist Eugénie Bastié (cited by Gauthey, 2025) speaks out against this "scientific imperialism." For her, if one limits oneself to scientifically established, objective statements, without activism, then freedom of expression is over. She claims the right to fake news, manipulation, and propaganda. This is all well and good in the general field of debate, but certainly not within the framework of a scientific debate.

According to geneticist Loic Yang (University of Queensland, Australia), cited by Kaiser (2023), creating a taboo on multiethnic genetic studies prevents scientific progress. According to behavioral geneticist Abdel Abdellaoui (Amsterdam University Medical Center), cited also by Kaiser (2023), such studies are inevitable, even if it is advisable to interpret them with caution and nuance.

General Conclusion

"Scientific antiracism" is an undertaking that has met with great success, both among specialists and the general public. It has imprinted on many minds the view that what is still called "races" in common parlance is a superficial phenomenon, one that ultimately holds little importance, which is negligible in the strong sense of the term: deserving of nothing more than to be neglected, set aside. For a time, biology seemed to corroborate this view. This was the "golden age of scientific antiracism."

However, this picture has been increasingly tarnished, on the one hand by a finer analysis of the scientific data from this "golden age"—particularly a questioning of the "smoking gun argument" of *F*st—and on the other hand, and above all, by the advent of genomics, which has allowed for a much finer characterization of human populations. It has become clear that scientific antiracism was largely based on promoting highly debatable definitions of human races (typological concept, phylogenetic concept, social science definitions partially or totally disconnected from biology). Indeed, these definitions in no way align with those of physical anthropology (Coon, 1962; Vallois, 1976) and population genetics (Ayala and Kiger, 1984; Dobzhansky, 1962a; Hartl, 1985; King and Stanfield, 1990; King et al., 2012; Mayr, 1963, 2002). In these debates about scientific antiracism, these definitions appear little or not at all.

In light of this new perspective, the point is not to bluntly state that "human races *do exist*" or that they "*are* subspecies/geographical races" (no more than one can assert the opposite of these two propositions). As repeated several times, we are not talking here about *absolute truth*, but only about *models*. The "Move along, nothing to see here" model, the one suggested, at least implicitly, by scientific antiracism, appears to have a defective predictive power. This means that the mantra: "human races do not exist" ceases to be an incontestable dogma and becomes a proposition that can and must be debated. Contrary to what has often been asserted (Graves et al., 2025), it is by no means the subject of a consensus.

Despite this weakening of scientific antiracism, many authors continue to hold this position, which still has undeniable success, particularly in the mainstream media (see for example Foucart, 2013).

Some authors propose a radical change in strategy in the face of this state of affairs. Thus, Laurent Alexandre (2018) does not hesitate to make the audacious proposal: "exceptionally, scholars must put scientific truth after the fundamental philosophical principle of the equality of all groups of men." This suggestion can be interpreted in two ways: either scholars stop treating scientific truth as an absolute requirement to defend the principle of equality, or these same scholars may "tinker"

DOI: 10.1201/9781003647393-10

scientific truth to serve equality. Stoczkowski (2006) stimulatingly titles his article: "Should antiracism break with science?" However, for this author, the point is not to neglect the scientific contribution but to separate a moral position, antiracism, from the scientific data. Doing the opposite would, in his view, lead to perverse effects such as discrimination in access to medical treatments. The Association of Black Cardiologists indeed supports the use of the ethno-specific drug BiDil. According to this association, an overrepresentation of individuals of European ancestry in biomedical studies leads to the development of treatments that primarily benefit white people. This aligns with the recommendations of Oni-Orisan et al. (2021).

I profoundly believe that this elimination of all ideological, moral or emotional bias upstream from the scientific process is an absolute necessity. Whatever the nobility of an ideology, in this case, antiracism, it should not influence the scientists' steps in their explorations and conclusions. This does not mean, of course, that the practice of science can occur without ethical concerns: in their daily work, researchers, like all citizens, must respect basic rules: have satisfactory relationships with their collaborators, do their best to train their students and find them career opportunities, banish everything called "scientific misconduct," such as "fudging results," or avoiding citing deserving colleagues in their publications, or citing them incorrectly, or relying on artificial intelligence to write their articles. But what must be absolutely banned is being influenced or led by personal beliefs to conduct research, desperately seeking data that will corroborate those beliefs, and concealing data that contradicts them. This is a difficult asceticism, as we all have cultural biases and we are all subject to influences, often unconscious. However, it seems to me that this approach is crucial, as the grip of ideology on science appears to be increasing, judging by the recent and debatable positions taken by major North American and French medical organizations. I have called this approach "reciprocal sanctuarization" (Tibayrenc, 2019). It has similarities with the so-called "NOMA" (non-overlapping magistera) approach, promoted in particular by the American paleontologist Stephen J. Gould, which dictates that science and religion are two domains of human thought with distinct goals, which should not interfere with each other, and which should respect each other.

Reciprocal sanctuarization must be particularly enforced in the present debate concerning human races and racism, a debate that urgently needs to be de-emotionalized due to its highly polemical nature. The variability of our species' geographical populations, *whatever name one gives them*, is an incomparable object of study, widely used: (https://drive.google.com/file/d/1qWNbPeyBCwDL9CgzoO MrHguFm9MeX4f2/view). (https://drive.google.com/file/d/1pffskqiM3WuPeD7gL IWYsSiXyTdwoeso/view) for analyzing the genetic control of important phenotypes, particularly diseases, and a valuable tool for developing personalized medicine. To obscure or deny this geographical variability in the name of a misunderstood antiracism is, in my humble opinion, inaccurate science and misled medicine. It will not make racism disappear and risk corrupting science and medicine.

References

Abdellaoui, A., Yengo, L., Verweij, K. J. H. and Visscher, P. M. 2023. 15 years of GWAS discovery: Realizing the promise. *Am. J. Hum. Genet.* 110: 179–194.

Adamon, D. 1949. The Seventy-Five Per Cent Rule for Subspecies. *The Condor* 51: 250–258. https://doi.org/10.2307/1364805.

Adleberg, J., Wardeh, A., Doo, F. X., Marinelli, B., Cook, T. et al. 2022. Predicting patient demographics from chest radiographs with deep learning. *J. Am. Coll. Radiol.* 19: 1151–1161.

Ahmadu, F. S. N., Bader, D., Boddy, J., Camara, M., Carver, N. et al. 2025. Harms of the current global anti-FGM campaign. *J. Med. Ethics* 0:1–8. doi:10.1136/jme-2025–110961

Aitken, C., Bosly, R., Bridgman, R., Cowley, A., Henry, S. A. C. et al. 1995. *Dictionary of Science and Technology*. Walker, P. M. B. Editor. Larousse, 95 Madison Avenue, New York.

Alcock, J. 2017. Human Sociobiology and Group Selection Theory. pp. 383–396. *In:* Michel Tibayrenc and Francisco J. Ayala. (eds.). *On Human Nature: Biology, Psychology, Ethics, Politics, and Religion*. San Diego : Elsevier/Academic Press.

Alexandre, L. 2013. "La guerre des cerveaux." (War of Brains) *Le Monde*, 7th March. https://www.lemonde.fr/sciences/article/2013/03/07/la-guerre-des-cerveaux_1844487_1650684.html

Alexandre, L. 2018. La génétique sur le terrain miné des "races" (Genetics on the minefield of "races"). *Le Monde*, 12th April. https://www.lemonde.fr/sciences/article/2018/04/12/la-genetique-sur-le-terrain-mine-des-races_5284322_1650684.html

Alexandre, L. 2024. *ChatGPT va nous rendre immortels*. (ChatGPT will make us immortal). JCLattès, Paris, France.

Ali-Khan, S. E., Krakowski, T., Tahir, R., and Daar, A. S. 2011. The use of race, ethnicity and ancestry in human genetic research. *Hugo J.* 5: 47–63.

Allentoft, M., Sikora, M., Fischer, A., Sjögren, K. G., Ingason, A. et al. 2024. 100 ancient genomes show repeated population turnovers in Neolithic Denmark. *Nature* 625: 329–337.

Alves, I., Srámkovaá Hanulová, A., Foll, M., and Excoffier, L. 2012. Genomic data reveal a complex making of humans. *PLoS Genet.* 8(7): e1002837. doi:10.1371/journal.pgen.1002837.

American Association of Anthropologists 1998. Statement on race. *Am. Anthropol.* 100: 712–713.

American Society of Human Genetics Board of Directors 2023. On the report of the ASHG "Facing Our History—Building an Equitable Future" initiative. *Am. J. Hum. Genet.* 110: 375–376.

Amuzu, S., Xie, A. X., Bai, X., Pekala, K. R., Pickersgill, N. A. et al. 2025. Meta-analysis reveals differences in somatic alterations by genetic ancestry across common cancers. *Nature Genet.* 67: 2655–2660.

Anonymous 2002a. Interview with Alan Goodman. *Race. The power of an illusion.* https://www.racepowerofanillusion.org/interviews/interview-alan-goodman-2002.

Anonymous 2002b. Interview with Richard Lewontin. *Race- the power of an illusion.* https://www.racepowerofanillusion.org/interviews/interview-richard-lewontin-2002.

Anonymous 2021a Controverse sur les rapports entre race et intelligence (Controversy about Relationships between Race and Intelligence). *Wikipedia,* 24th November. https://fr.wikipedia.org/w/index.php?title=Controverse_sur_les_rapports_entre_race_et_intelligence&oldid=188279041.

Anonymous 2021b race and genetics; *Wikipedia* 14th December. https://en.wikipedia.org/w/index.php?title=Race_and_genetics&oldid=1060242543.

Anonymous 2022a. Race humaine (Human race). *Wikipedia* 27th November. https://fr.wikipedia.org/wiki/Race_humaine?

Anonymous 2022b. Scientific racism. *Wikipedia* 7th February. https://en.wikipedia.org/w/index.php?title=Scientific_racism&oldid=1070382126

Anonymous 2022c. Fits and starts: anti-racism efforts in the biosciences. *Nature* 612: 177–179.

Anonymous 2023. IQ and the wealth of nations. *Wikipedia*, 10th August. https://fr.wikipedia.org/wiki/IQ_and_the_Wealth_of_Nations#.

Anonymous 2024a. Race (human categorization). *Wikipedia* 3rd May. https://en.wikipedia.org/wiki/Race_(human_categorization).

Anonymous 2024b. Race and intelligence. *Wikipedia* 7th December. https://en.wikipedia.org/wiki/Race_and_intelligence.

Anonymous 2024c. Hybrid (biology). *Wikipedia*, 15th December. https://en.wikipedia.org/wiki/Hybrid_(biology).

Anonymous 2025. We need to talk about human genome editing. *Nature* 637 : 252.

Arama, C., Maiga, B., Dolo, A., Kouriba, B., Traoré, B. et al. 2015. Ethnic differences in susceptibility to malaria: What have we learned from immuno-epidemiological studies in West Africa?" *Acta tropica* 146: 152–56.

Armengol, L., Villatoro, S., González, J. R., Pantano, L., Garcıá-Aragonés, M. et al. 2009. Identification of Copy Number Variants Defining Genomic Differences among Major Human Groups. *PLoS ONE* 4(9): e7230. doi:10.1371/journal.pone.0007230.

Arpone, M., Turbitt, E., and MvEven, A. 2024. Race, ethnicity, and ancestry reporting in genetic counseling research: A focused mapping review and synthesis. *JoGC*: e1884. https://doi.org/10.1002/jgc4.1884

Arvey, R. D., Bouchard, Jr., T. J., Kaufman, N. L., Carroll, J. B., Cattell, R. B. et al. 1994. Mainstream Science on Intelligence. *The Wall Street Journal*, December 13.

ASHG perspective 2018. ASHG Denounces Attempts to Link Genetics and Racial Supremacy. *Am. J Human Genet.* 103: 636.

Atkinson, E. G., Audesse, A. J., Palacios, J. A., Bobo, D. M., Webb, A. E. et al. 2018. No Evidence for Recent Selection at FOXP2 among Diverse Human Populations. *Cell* 174: 1424–1435.e15 https://doi.org/10.1016/j.cell.2018.06.048.

Attwaters, M. 2024. Ancient migration and the modern genome. *Nature Rev. Genet*. 25: 159–164.

Austad, S. N. and Finch, C. E. 2017. Human life history evolution: new perspectives on body and brain growth. pp. 221–34. *In*: Michel Tibayrenc and Francisco J. Ayala (eds.). *On Human Nature: Biology, Psychology, Ethics, Politics, and Religion*. San Diego : Elsevier/Academic Press.

Avise, J. C. 2004. *Molecular Markers, Natural History and Evolution, 2nd* edn. New York and London: Chapman and Hall.

Ayala, F. J., and Kiger, J. A. Jr. 1984. *Modern genetics* (Second edition). Benjamin/Cummings, Menlo Park and London.

Ayub, Q., Yngvadottir, B., Chen, Y., Xue,Y., Hu, M. et al. *FOXP2* Targets Show Evidence of Positive Selection in European Populations. *Am. J. Human Genet.* 92: 696–706

Badeau, K. 2023. Pourquoi la notion de racisme anti-Blancs fait-elle débat ? (Why is the concept of anti-white racism a subject of debate?) *Le Point,* 13th December. https://www.lepoint.fr/postillon/pourquoi-la-notion-de-racisme-anti-blancs-fait-elle-debat-13-12-2023-2546869_3961.php?M_BT=20588364922&boc=208785&nl_key=1a4dad71824665c5a198b17610c4f0a226f370549e46ba bfbc4da61c2f46680b#xtor=EPR-6-[Newsletter-Mi-journee]-20231213-[Debat_1].

Badinter, E. 1980. *L'amour en plus* (Love, Plus). Flammarion, Paris, France.

Badinter, E. 2003. *Fausse route* (Wrong Way). Editions Odile Jakob, Paris, France.

Balter, M. 2015. Mysterious Indo-European homeland may have been in the steppes of Ukraine and Russia. *Science*, 13 février. https://www.science.org/content/article/mysterious-indo-european-homeland-may-have-been-steppes-ukraine-and-russia

Banda, Y., Kvale, M. N., Hoffmann, T. J., Hesselson, S. E., Ranatunga, D. et al. 2015. Characterizing Race/Ethnicity and Genetic Ancestry for 100,000 Subjects in the Genetic Epidemiology Research on Adult Health and Aging (GERA) Cohort. *Genetics* 200 : 1285–1295.

Bantignies, F., and Giacomo, C. 2025. Nous héritons de la génétique de nos parents, mais quid de l'épigénétique ? (We inherit genetics from our parents, but what about epigenetics?). *The Conversation*, 30th March. https://theconversation.com/nous-heritons-de-la-genetique-de-nos-parents-mais-quid-de-lepigenetique-247948?utm_medium=email&utm_campaign=La%20lettre%20de%20The%20Conversation%20France%20du%2030%20mars%202025%20-%20

3317133876&utm_content=La%20lettre%20de%20The%20Conversation%20France%20du%20 30%20mars%202025%20-%203317133876+CID_d9546e1a64307710f8e86e57ec164806&utm_source=campaign_monitor_fr&utm_term=Nous%20hritons%20de%20la%20gntique%20de%20 nos%20parents%20mais%20quid%20de%20lpigntique.

Barbujani, G., and Colonna, V. 2010. Human genome diversity: frequently asked question. *Trends Genet.* 26: 285–95.

Belbin, G. M., Cullina, S., Wenric, S., Soper, E. R., Glicksberg, B. S. et al. 2021. Toward a fine-scale population health monitoring system. *Cell*, 184: 2068–2083.

Belizário, J. E. 2013. The humankind genome: from genetic diversity to the origin of human diseases *Genome* 56: 705–716.

Ben-Ari, Y., Clerget-Darpoux, F., and Danchin, E. 2023. Génomique : la part de variabilité liée à l'environnement ne peut plus être niée (Genomics: the degree of variability linked to the environment can no longer be denied). *Le Monde*, 13th May. https://www.lemonde.fr/sciences/article/2023/05/12/genomique-la-part-de-variabilite-liee-a-l-environnement-ne-peut-plus-etre-niee_6173078_1650684.html.

Benjamin, K. J. M., Chen, Q., Eagles, N. J., Huuki-Myers, L. A., Collado-Torres, L. et al. 2024. Analysis of gene expression in the postmortem brain of neurotypical Black Americans reveals contributions of genetic ancestry. *Nature Neurosci.* https://www.nature.com/articles/s41593-024-01636-.

Berent, I. 2022. Pourquoi la science de la nature humaine n'est pas à craindre (Why the science of "human nature" is not to be feared). *Le Point*, 11th December. https://www.lepoint.fr/debats/pourquoi-la-science-de-la-nature-humaine-n-est-pas-a-craindre-11-12-2022-2501281_2.php?M_BT=20588364922#xtor=EPR-6-[Newsletter-Mi-journee]-20221211-[Debat_2].

Bergström, A., McCarthy, S. A., Hui R., Almarri, M. A., Ayub, Q. et al. 2020. Insights into human genetic variation and population history from 929 diverse genomes. *Science* 367: 1–10.

Bernichan J. 2023. Le QI est-il héréditaire? (Is IQ hereditary ?) *Yahoo ! actualités* 22th May. https://fr.news.yahoo.com/le-qi-est-il-hereditaire-081612899.html.

Berry, A., and Petrov, D. A. 2021. Richard C. Lewontin (1929–2021). Groundbreaking evolutionary geneticist. *Science* 373: 745.

Berryessa, C. M., and Cho, M. K. 2013. Ethical, Legal, Social, and Policy Implications of Behavioral Genetics. *Annual Rev. Genomics Human Genet.* 14: 515–34.

Bettayeb, K. 2025. Afrique, berceau du genre *Homo*. (Africa, craddle of the genus *Homo*) *J. CNRS*. 28th July. https://lejournal.cnrs.fr/articles/lafrique-berceau-du-genre-homo.

Betti, M. J., Jaworski, J., Zhao, S., Rao, J. S., Ryan, B. M. et al. 2025. Genetic analysis in African ancestry populations reveals genetic contributors to lung cancer susceptibility. *Am. J. Human Genet.* 112: 2102–2114.

Bherer, M. O. 2019. Le retour controversé de la race dans les sciences sociales (The controversial return of "race" in the social sciences). *Le Monde*, 13th August. https://www.lemonde.fr/idees/article/2019/07/12/le-retour-controverse-de-la-race-dans-les-sciences-sociales_5488430_3232.html.

Bick, A. G., and the All of Us Research Program Genomics Investigators. 2024. Genomic data in the All of Us Research Program. *Nature*, DOI https://doi.org/10.1038/s41586-023-06957-x.

Bien, S. A., Wojcik, G. L., Hodonsky, C. J., Gignoux, C. R., Cheng, I. et al. 2019. The Future of Genomic Studies Must Be Globally Representative: Perspectives from PAGE. *Annu. Rev. Genom. Hum. Genet.* 20: 181–200.

Bikfalvi, A., and Kuntz, M. 2020. L'inquiétante campagne de racialisation de la science (the Disturbing Campaign of Science Racialization). *Le Point,* 19th September. https://www.lepoint.fr/debats/l-inquietante-campagne-de-racialisation-de-la-science-19-09-2020-2392654_2.php.

Bindon, J. R. 2020. Race in the wake of the Human Genome Project. Preprint DOI: 10.13140/RG.2.2.33269.40162.

Birken, M. 2025. Les connaissances sur *Homo sapiens* bouleversées par cette étude d'un crâne vieux d'un million d'années. (Knowledge of *Homo sapiens* revolutionized by this study of a million-year-old skull) *Huffington Post*, 26th September. https://www.huffingtonpost.fr/science/article/les-connaissances-sur-homo-sapiens-bouleversees-par-cette-etude-d-un-crane-vieux-d-un-million-d-annees_255289.html.

Bloch-Lainé, V. 2022. Charles King: Le fondateur de l'anthropologie culturelle Franz Boas considérait l'humanité comme indivisible (Charles King "The founder of cultural anthropology, Franz Boas, considered humanity to be indivisible."). *Libération* : 11th March. https://www.liberation.fr/idees-et-debats/charles-king-le-fondateur-de-lanthropologie-culturelle-franz-boas-considerait-lhumanite-comme-indivisible-20220311_A57O3FIKCBAEBOVDYUP57MYP3U/.

Blodgett, S. L., and Talat, Z. 2024. AI responds with racism to African American English. *Nature* 633: 40–41.

Boahen, C. K., Temba, G. S., Kullaya, V. I., Matzaraki, V., Joosten, L. A. B. et al. 2022. A functional genomics approach in Tanzanian population identifies distinct genetic regulators of cytokine production compared to European population. *Am. J. Human Genet.* 109: 471–485.

Boas, F. 1912. Changes in the bodily form of descendants of immigrants. *Amer. Anthropol.* 14: 530–62.

Bolognini, D., Halgren, A., Lou, R. N., Raveane, A., Rocha, J. L. et al. 2024. Recurrent evolution and selection shape structural diversity at the amylase locus. *Nature* 634: 617–625.

Bon, C. 2025. Des nouvelles du passé. Histoires de paléogénétique. Quand Néandertal rencontre *Sapiens* : ce que l'on sait, et ce que l'on ignore encore du métissage entre Néandertal et *Homo sapiens* (News from the past. Tales of paleogenetics. When Neanderthals met *Homo sapiens*: what we know, and what we still don't know, about interbreeding between Neanderthals and *Homo sapiens*). *Médecine/Sciences* 41 : 386–391.

Bonniol, J. L., Edwards-Grossi, E., and Wang, S. 2021. Introduction au dossier Race et biologie (Introduction to the file 'Race and biology'). *Cahiers de l'Urmis* https://journals.openedition.org/urmis/2278.

Boqueho, V. 2023. D'où venons-nous ? Les races existent-elles ? (Where do we come from? Do races exist?) *Herodote*, 5th July. https://www.herodote.net/Les_races_existent_elles_-synthese-3327-554.php.

Bordenave, V. 2018a. Le premier Européen, *Homo antecessor*, a au moins 800 000 ans. (The first European, *Homo antecessor*, is at least 800,000 years old.) *Le Figaro*, 27th June. https://www.lefigaro.fr/sciences/2018/06/26/01008-20180626ARTFIG00260-le-premier-europeen-homo-antecessor-a-au-moins-800000-ans.php

Bordenave, V. 2018b. Cheddar Man' : l'ancêtre des Britanniques avait-il la peau noire ? (Cheddar Man': was the ancestor of the British black-skinned?) *Le Figaro*, 14th February. https://www.lefigaro.fr/sciences/2018/02/13/01008-20180213ARTFIG00319-cheddar-man-l-ancetre-des-britanniques-avait-il-la-peau-noire.php

Bordenave, V. 2025. Il y a cinq mille ans, la conquête éclair de l'Europe par les Yamnayas. (Five thousand years ago, the Yamnayas' lightning conquest of Europe) *Le Figaro*, 7th February. https://www.lefigaro.fr/sciences/il-y-a-cinq-mille-ans-la-conquete-eclair-de-l-europe-par-les-yamnayas-20250207.

Bouchard Jr, T.J. 1998. Genetic and environmental influences on adult intelligence and special mental abilities. *Human Biol.* 70: 257–79.

Boule, M. 1920. *Les hommes fossiles - Éléments de paléontologie humaine* (Fossil Men - Elements of Human Paleontology). Paris, France : Masson.

Bousenna, Y. 2024. Pierre Gaussens, sociologue : Les études décoloniales réduisent l'Occident à un ectoplasme destructeur (Pierre Gaussens, sociologist: Decolonial studies reduce the West to a destructive ectoplasm.). *Le Monde*, 24th November. https://www.lemonde.fr/idees/article/2024/11/24/pierre-gaussens-sociologue-les-etudes-decoloniales-reduisent-l-occident-a-un-ectoplasme-destructeur_6412217_3232.html.

Bouts, P. 2019. Le peuplement de l'Europe ; La révélation de la paléogénétique et les Indo-Européens (The peopling of Europe; The revelation of paleogenetics and the Indo-Europeans). *Nouvelle Ecole* 68 : 9–50.

Boye, C., Nirmalan, S., Ranjbaran, A., and Luca, F. 2024. Genotype × environment interactions in gene regulation and complex traits. *Nature Genet.* 56 : 1057–1068.

Brafman, N. 2024. La drépanocytose, une anémie d'origine génétique, enfin dépistée chez tous les nouveau-nés (Sickle cell disease, an anemia of genetic origin, is finally being screened for in all newborns). *Le Monde*, 1st November. https://www.lemonde.fr/sciences/article/2024/11/01/la-drepanocytose-une-anemie-d-origine-genetique-enfin-depistee-chez-tous-les-nouveau-nes_6369966_1650684.html?search-type=classic&ise_click_rank=2.

Brett, A. S., and Goodman, C. W. 2021. First Impressions — Should We Include Race or Ethnicity at the Beginning of Clinical Case Presentations? *New Engl. J. Med.* 385: 2497–2499.

Brothers, K. B., Bennett, R. L., and Cho, M. K. 2021. Taking an antiracist posture in scientific publications in human genetics and genomics. *Genetics in Medicine* https://doi.org/10.1038/s41436-021-01109-w.

Brown, G. R., H. Barrett, C., and Kevin N., Lala, K. L. 2025. Revisiting the human sociobiology debate. What have we learned 50 years on? *Science* 389: 580–581.

Bucheton, B., Kheir, M. M., El-Safi, S. H., Hammad, A., Mergani, A. et al. 2002. The interplay between environmental and host factors during an outbreak of visceral leishmaniasis in eastern Sudan. *Microbes Infect.* 4 : 1449–1457.

Budowle, B., and Sajantila, A. 2024. Short tandem repeats—how microsatellites became the currency of forensic genetics. *Nature Rev. Genet.* 25: 450.

Burchard, E. G., Oh, S. S., Foreman, M. G., and Celedón, J. C. 2015. Moving toward True Inclusion of Racial/Ethnic Minorities in Federally Funded Studies. A Key Step for Achieving Respiratory Health Equality in the United States. *Am. J. Respir. Crit. Care Med.* 191 : 514–521.

Burdick, K. E., Lencz, T., Funke, B., Finn, C.T., Szeszko, P. R. et al. 2006. Genetic variation in DTNBP1 influences general cognitive ability. *Human Molec. Genet.* 15 : 1563–1568.

Byappanahalli, A., Omoniyi, V., Hooten, N. N., Smith, J. T., Mode, N. A. et al. 2024. Extracellular vesicle mitochondrial DNA levels are associated with race and mitochondrial DNA haplogroup. *iScience* 27: 108724.

Byeon, Y. J. J., Islamaj, R., Yeganova, L., Wilbur, W. J., Lu, Z. et al. 2021. Evolving use of ancestry, ethnicity, and race in genetics research—A survey spanning seven decades. *Am. J. Hum. Genet.* 108-2215-2223.

Cailloce, L. 2022. Les nouvelles frontières du vivant (the new frontiers of life). *J. CNRS*, 28th July. https://lejournal.cnrs.fr/articles/les-nouvelles-frontieres-du-vivant.

Callaway, E. 2018. The battle for common ground. *Nature* 555 : 573–76. doi : 10.1038/d41586-018-03773-6.

Callaway, E. 2021. Oldest human DNA reveals recent Neanderthal mixing. *Nature* 592: 339.

Callaway, E. 2025. How one language family took over the world: ancient DNA traces its spread. *Nature*: https://doi.org/10.1038/d41586-025-00382-y.

Carlson, J., Henn, B. M., Al-Hindi, D. R., and Ramachandran, S. 2022. Counter the weaponization of genetics research by extremists. *Nature* 610: 444–447.

Carmelli, D., and Cavalli-Sforza, L. L. 1979. The Genetic Origin of the Jews: A Multivariate Approach. *Human Biol.* 51: 41–61.

Caulfield, T., Fullerton, S. M., Ali-Khan, S. E., Arbour, L., Burchard, E. G. et al. 2009. Race and ancestry in biomedical research: exploring the challenges. *Genome Med.* 1: 8.

Cavalli-Sforza, L. L. 2005. The Human Genome Diversity Project: past, present and future. *Nature Rev. Genet.* 6: 333–340.

Cavalli-Sforza, L. 2007. Human evolution and its relevance for scientific epidemiology. *Annnu. Rev. Genomics Hum. Genet.* 8 : 1–15.

Cavalli-Sforza, L. L., and Cavalli-Sforza, F.. 1997. *Qui sommes-nous? Une histoire de la diversité humaine* (Who are we? A history of human diversity). Paris, France : Flammarion.

Cavalli-Sforza, L. L., and Feldman, M. 2003. The Application of Molecular Genetic Approaches to the Study of Human Evolution. *Nature Genetics* 33 Suppl (Suppl): 266–75.

Ceci, S., and Williams, W. M. 2009. Should scientists study race and IQ? YES: The scientific truth must be pursued. *Nature* 457: 788–789.

Cerdeña, J. P., Grubbs, V., and Non, A. L. 2022. Genomic supremacy: the harm of conflating genetic ancestry and race. *Human Genomics* 16: 18.

Chapman, S. J., and Hill, A. V. S. 2012. Human genetic susceptibility to infectious disease. *Nature Rev. Genet.* 13 : 175–188.

Chavet, S. 2020. Thomas Chatterton Williams, Glenn Loury, Coleman Hughes… Les anticonformistes de l'antiracisme (Thomas Chatterton Williams, Glenn Loury, Coleman Hughes… the nonconformists of antiracism). *Le Monde*, 18th September. https://www.lemonde.fr/m-le-mag/article/2020/09/18/thomas-chatterton-williams-glenn-loury-coleman-hughes-les-anticonformistes-de-l-antiracisme_6052654_4500055.html.

Chemin, A. 2022. Derrière le débat sur le wokisme, les trois mutations du racisme : biologique, culturel, systémique (Behind the debate on "wokism" lie the three mutations of racism: biological, cultural, systemic). *Le Monde*, 6th January. https://www.lemonde.fr/idees/article/2022/01/06/biologique-culturel-systemique-le-racisme-en-trois-mutations_6108458_3232.html.

Chen, C. L., Gold, G. J., Cannesson, M., and Lucero, J. M. 2021. Calling Out Aversive Racism in Academic Medicine. *N. Engl. J. Med.* 27: 2499–2501.

Chen, Y., Liu, S., Ren, Z., Wang, F., Liang, Q. et al. 2024. Cross-ancestry analysis of brain QTLs enhances interpretation of schizophrenia genome-wide association studies. *Am. J. Human Genet.* 111: 2444-2457.

Cheng, S. J., Gaynor, K. M., Moore, A. C., Darragh, K., Estien, C. O. et al. 2023. Championing inclusive terminology in ecology and evolution. *Trends Ecol. Evol.* 38: 381–384.

Chimpanzee Sequencing and Analysis Consortium 2005. Initial sequence of the chimpanzee genome and comparison with the human genome. *Nature* 437 : 69–87.

Chneiweiss, H. 2019. Éléments pour une éthique de la communication des faits scientifiques (Elements for an ethics of communicating scientific facts). *Médecine Sciences* 35: 95–96.

Cho, M. K., and Sankar, P. 2004. Forensic medicine and ethical, legal and social implications beyond the clinic. *Nature Genet. Suppl.* 38: S8–S11.

Choudhury, A. R., and Singh, K.K. 2017. Mitochondrial determinants of cancer health disparities. *Semin Cancer Biol.* 47: 125–146.

Clerget-Darpoux, F., and Genin, E. 2023. Notre ADN détermine-t-il vraiment notre intelligence et nos risques de maladies ? (Does our DNA really determine our intelligence and disease risk?) *The Conversation.* https://theconversation.com/notre-adn-determine-t-il-vraiment-notre-intelligence-et-nos-risques-de-maladies-198080.

Cochran, G., and Harpending, H. 2009. *The 10,000 year explosion- How civilization accelerated human evolution.* New York: Basic books. A member of the Perseus Books Group.

Collins, F. S. 2004. What do we do and don't know about "race", "ethnicity", genetics and heath at the dawn of the genome era." *Nature Genet. Suppl.* 36: S13–S15.

Collins, R. L., and Talkowski, M. E. 2025. Diversity and consequences of structural variation in the human genome. *Nature Rev. Genet.* 26: 443–462.

Combis, H. 2018. Le concept de race peut-il s'appliquer à l'espèce humaine ? (Can the concept of "race" be applied to the human species?) *France culture*, 12th July. https://www.franceculture.fr/sciences/le-concept-de-race-peut-il-sappliquer-lespece-humaine.

Condemi, S., and Savatier, F. 2016. *Neandertal, mon frère* (Neandertal, my Brother). Flamamrion éd. Paris, France.

Conrad, D. F., and Hurles, M. E. 2007. The population genetics of structural variation. *Nature Genet. Suppl.* 39: 530–36.

Coon, C. S. 1962. *The origin of races*. New York: Alfred A. Knoff.

Cooper, R. S., Jay S., Kaufman, J.S., and Ward, R. 2003. Race and Genomics. *New England J. Med.* 348: 1166–1170.

Corona, E., Chen, R., Sikora, M., Morgan, A. A., Patel, C.J. et al. 2013. Analysis of the Genetic Basis of Disease in the Context of Worldwide Human Relationships and Migration. *Plos Genet.* 9(5): e1003447. doi:10.1371/journal.pgen.1003447.

Corpas, M., Guio, H., Lopez-Correa, C., and Fatumo, S. 2025. Why genomic diversity should not be framed by census alone. *Nature Genet.* 57: 1793–1794.

Coyne, J. A., and Maroja, L. S. 2023. The Ideological Subversion of Biology. *Skeptical Inquirer*. Volume 47, N°4; July/August. https://skepticalinquirer.org/2023/06/the-ideological-subversion-of-biology/

Cracraft, J. 1983. *Species concept and speciation analysis*. pp. 159–187. *In*: Johnson, R. F. (ed.). Current ornithology. Plenum Press, New York.

Crawford, N. G., Kelly, D. E., Hansen, M. E. B., Beltrame, M. H., Fan, S. et al. 2017. Loci associated with skin pigmentation identified in African populations. *Science* 358: eaan8433.

Cross-Disorder Group of the Psychiatric Genomics Consortium. 2013. Identification of risk loci with shared effects on five major psychiatric disorders: a genome-wide analysis. *The Lancet* 381: 1371–1379.

Currat, M., and Excoffier L. 2004. Modern humans did not admix with Neanderthals during their range expansion into Europe. *Plos Biol.* 2(12): e421.

Curry, A. 2024. Mystery woman's DNA reveals close family ties between Europe's earliest people. *Science*, News, 12th December. https://www-science-org.inee.bib.cnrs.fr/content/article/mystery-woman-s-dna-reveals-close-family-ties-between-europe-s-earliest-people.

Daco, P. 1973. *Les Prodigieuses Victoires de la psychologie* (The Prodigious Achievements of Psychology). Paris, France : Éditions Marabout.

Dean, M., Carrington, M., Winckler, C., Huttley, G. A., Smith, M.W. et al. 1996. Genetic restriction of HIV-1 infection and progression to AIDS by a deletion allele of the CKR5 structural gene. *Science* 273 : 1856–1862.

De Beauvoir, S. 1949. *Le deuxième sexe* (the Second Sex). Gallimard, Paris, France.

de Benoist, A. 1979. *Les idées à l'endroit* (ideas in place). Éditions libres Hallier, Paris, France.

de Candia, T. R., Hong Lee, S., Yang, J., Browning, B. L., Gejman,, P. V. et al. 2013. Additive Genetic Variation in Schizophrenia Risk Is Shared by Populations of African and European Descent. *Am. J. Hum. Genet.* 93: 463–470.

de Tocqueville, A. 1835. De la démocratie en Amérique (Democraty in America). Paris, France, librairie de Charles Gosselin.

Delgado, C., Baweja, M., Crews, D. C., Eneanya, N. D., Gadegbeku, C. A. et al. 2021. A Unifying Approach for GFR Estimation: Recommendations of the NKF-ASN Task Force on Reassessing the Inclusion of Race in Diagnosing Kidney Disease. *Am. J. Kidney Dis.* 79(2): 268–288.

Demoule, J. P. 2014. *Mais où sont passés les Indo-Européens?* (But where did the Indo-Europeans go?) La librairie du XXIe siècle. Paris, France : Éditions du Seuil.

Deyrup, A., and Graves, J. L. 2022. Racial Biology and Medical Misconceptions. *New Engl. J. Med.* 386: 501–503.

DiAngelo, R. 2011. *White Fragility. Int. J. Crit. Pedagogy* 3: 54–70.

Dietrich, M. R. 2021. Richard C. Lewontin (1929–2021). Pioneer of molecular evolution who campaigned against biological racism. *Nature* 595: 489.

Dignat, A. 2018. *Homo sapiens* convole avec Neandertal (*Homo sapiens* mates with Neanderthal). *Les Amis d'Herodote*. https://www.herodote.net/histoire/evenement.php?jour=20100507&ID_dossier=250

Dobzhansky, T. 1937. *Genetics and the origin of species*. New York: Columbia University Press.

Dobzhansky, T. 1962a. *Mankind Evolving. The Evolution of the Human Species*. New Haven, CT: Yale University Press.

Dobzhansky, T. 1962b. Comment on Frank B. Livingstone: "on the non-existence of human races". *Curr. Anthropol.* 3: 279–281.

Dobzhansky, T. 1964. *Heredity and the Nature of Man*. Harcourt, Brace & World Inc., New York

Dobzhansky, T. 1973. *Genetic Diversity and Human Equality*. Basic Book, Inc., Publishers.

Dobzhansky, T. and Spassky, B. 1959. *Drosophila paulistorum*, a cluster of species in statu nascendi. *Proc. Natl. Acad. Sci. USA* 45: 419–428.

Donnelly, B., Berrang-Ford, L., Labbé, J., Twesigomwe, S., Lwasa, S. et al. 2016. *Plasmodium falciparum* malaria parasitaemia among indigenous Batwa and non-indigenous communities of Kanungu district, Uganda. *Malaria J.* 15: 254. DOI 10.1186/s12936-016-1299-1.

Donovan, B. M., Weindling, M., Amemiya, J., Salazar, B., Lee, D. et al. 2024. Humane genomics education can reduce racism. *Science* 383: 818–822.

Doron, C.O. 2013. Race et médecine. Une vieille histoire (Race and Medicine. An Old Story). *Médecine Sciences* 10: 919–922.

Dos Santos, G., and Lewino, F. 2021. Sans nul doute, le métissage est un accélérateur de l'évolution humaine (Without a doubt, interbreeding is an accelerator of human evolution). *Le Point*, 6th October. https://www.lepoint.fr/sante/sans-nul-doute-le-metissage-est-un-accelerateur-de-l-evolution-humaine-06-10-2021-2446505_40.php?M_BT=20588364922#xtor=EPR-6-[Newsletter-Mi-journee]-20211006-[Article_8].

Dos Santos, G., Lewino, F., and Tourbe, C. 2021. Hérédité : la fin de la tyrannie des gènes (Heredity: The End of the Tyranny of Genes). *Le Point*, 4th October. https://www.lepoint.fr/sante/heredite-la-fin-de-la-tyrannie-des-genes-04-10-2021-2446151_40.php?M_BT=20588364922#xtor=EPR-6-[Newsletter-Matinale]-20211005-[Article_7].

Droit, R. P. 2021. L'Antiracisme devenu fou , de Pierre-André Taguieff : la chronique "essais" de Roger-Pol Droit ("Antiracism gone mad", by Pierre-André Taguieff: Roger-Pol Droit's "essays" column.).

Le Monde, 19th November. https://www.lemonde.fr/livres/article/2021/11/19/l-antiracisme-devenu-fou-de-pierre-andre-taguieff-la-chronique-essais-de-roger-pol-droit_6102671_3260.html

Duncan, R. G., Krishnamoorthy, R., Harms, U., Haskel-Ittah, M., Kampourakis, K. et al. 2024. The sociopolitical in human genetics education. *Science* 383: 826–828.

Dynesen, H. B. 2018. A Striking Similarity: The Revolutionary Findings of Twin Studies. *Quillette*, 9th August. https://quillette.com/2018/08/09/a-striking-similarity-the-revolutionary-findings-of-twin-studies/

Edgar, H. J. H. 2009. Biohistorical Approaches to "Race" in the United States: Biological Distances Among African Americans, European Americans, and Their Ancestors. *Am. J. Phys. Anthropol.* 139: 58–67.

Edgar, H. J. H., and Hunley, K. L. 2009. Race Reconciled? How Biological Anthropologists View Human Variation. *Am. J. Phys. Anthropol*. 139: 1–4.

Editorial. 2018. Use and abuse of ancient DNA. *Nature* 555: 559.

Editorial 2000. Census, race and science. *Nature Genet*. 24: 97–98.

Editorial 2022a. Science must respect the dignity and rights of all humans. *Nature Human Behav*. 6: 1029–1031.

Editorial 2022b. Thoughts on a legacy. *Nature Rev. Genet*. 23: 385.

Edwards, A.W. 2003. Human genetic diversity: Lewontin's fallacy. *BioEssays* 25: 798–801.

Else, J., and Perkel, J. M. 2022. The giant plan to track diversity in research journals. *Nature* 602: 566–570.

Equy, L. 2018. Les députés votent la suppression du mot 'race' de la Constitution. (Deputies vote to remove the word 'race' from the Constitution) *Libération*, 27th June. https://www.liberation.fr/politiques/2018/06/27/les-deputes-votent-la-suppression-du-mot-race-de-la-constitution_1662341/.

Esoh, K., Yalcouyé, A., and Wonkam, A. 2025. Parasite-resistant variants in African genomes: implications for global genetic medicine. *Trends Parasitol*. 41: 536–546.

Ey, H., Bernard, P., and Brisset, C. 1960. *Manuel de psychiatrie* (Manual of Psychiatry). Masson Ed., Paris, France.

Fabre, M. 1999. *Situations-problèmes et savoir scolaire (Problem situations and academic knowledge)*. Presses Universitaires de France, Paris, France.

Farrer, L. A., Cupples, L. A., Haines, J. L., Hyman, B., Kukull, W.A. et al. 1997. Effects of age, sex, and ethnicity on the association between apolipoprotein E genotype and Alzheimer disease. A meta-analysis. APOE and Alzheimer Disease Meta Analysis Consortium. *JAMA* 278: 1349–1356.

Feng, X., Yin, Q., Gao, F., Lu, D., Fang, Q., Feng, Y. et al. 2025. The phylogenetic position of the Yunxian cranium elucidates the origin of *Homo longi* and the Denisovans. *Science* 389: 1320–1324.

Fernández-Crespo, T., Snoeck, C., Ordoño, J., Lefranc, P., Perrin, B. et al. 2025. Multi-isotope biographies and identities of victims of martial victory celebrations in Neolithic Europe. *Sci. Adv*. 11, eadv3162

Field, Y., Boyle, E. A., Telis, N., Gao, Z., Gaulton, K. J. et al. 2016. Detection of human adaptation during the past 2000 years. *Science* 354: 760–64.

Flanagin, A., Frey, T., Stacy L Christiansen, S. L., and AMA Manual of Style Committee 2021. Updated Guidance on the Reporting of Race and Ethnicity in Medical and Science Journals. *JAMA* Aug 17;326(7): 621–627. doi: 10.1001/jama.2021.13304.

Foucart, S. 2009. Sur la piste controversée des Indo-Européens (On the controversial trail of the Indo-Europeans). *Le Monde*, 19th June. https://www.lemonde.fr/planete/article/2009/06/19/sur-la-piste-controversee-des-indo-europeens_1208911_3244.html.

Foucart, S. 2013. Les "races humaines" existent ? Alors énumérez-les ! (Do Human Races Exist? Then list Them!) *Le Monde*, 3rd June. https://www.lemonde.fr/idees/article/2013/06/03/les-races-humaines-existent-alors-enumerez-les_3422652_3232.html#2jveJO13VrxdZL4J.99.

Foucart, S. 2023. En biologie, les "bons" et "mauvais" gènes font un inquiétant retour, alimentant les théories racialistes (In biology, "good" and "bad" genes are making a worrying comeback, fueling racialist theories). *Le Monde*, 20th January. https://www.lemonde.fr/idees/article/2023/01/20/en-biologie-les-bons-et-mauvais-genes-font-un-inquietant-retour-alimentant-les-theories-racialistes_6158611_3232.html.

Foucart, S. 2025. Le racialisme biologique revient aux États-Unis, mais aussi en France (Biological racialism returns to the United States, but also to France). *Le Monde*, 23rd September. https://www.lemonde.fr/idees/article/2025/09/23/le-racialisme-biologique-revient-aux-etats-unis-mais-aussi-en-france_6642591_3232.html.

Frances, A. 2013. *Saving Normal: An Insider's Revolt Against Out-of-Control Psychiatric Diagnosis, -5, Big Pharma, and the Medicalization of Ordinary Life.* New York: William Morrow.

Frost, P. 2019. The original industrial revolution. Did cold winters select for cognitive ability? *Psych.* 1: 166–181.

Fu, J, Zhang, Q., Wang, J., Wang, M., Zhang, B. et al. 2024. Cross-ancestry genome-wide association studies of brain imaging phenotypes. *Nature Genet.* 56: 1110–1120.

Fu, W., and Akey, J. M.. 2013. Selection and Adaptation in the Human Genome. *Ann. Rev. Genomics Human Genet.* 14: 467–89.

Fu, W., O'Connor, T. D., Jun, G., Kang, H. M., Abecasis, G. et al. 2013. Analysis of 6,515 exomes reveals the recent origin of most human protein-coding variants. *Nature* 493: 216–20.

Fumagalli, M., and Balloux, F. 2017. Natural selection associated with infectious diseases. pp. 177–91. *In*: Michel Tibayrenc and Francisco J. Ayala (eds.). *On Human Nature: Biology, Psychology, Ethics, Politics, and Religion.* San Diego : Elsevier/Academic Press.

Gannett, L. 2013. Theodosius Dobzhansky and the genetic race concept. *Studies History Philos. Biol. Biomed. Sci.* 44: 250–261.

Garcia-Pérez, R., Ramirez, J. M., Ripoll-Cladellas, A., Ferreira P. G., Ardlie, K. G. et al. 2023. The landscape of expression and alternative splicing variation across human traits. *Cell Genomics* 3: 100244.

Gaskin, D. J. 2025. Correcting the record on race and medicine. *Science* 390: 341.

Gaudieri, S., and John, M. 2017. Evolution of the Human Leukocyte Antigen System. In *On Human Nature: Biology, Psychology, Ethics, Politics, and Religion* edited by Michel Tibayrenc and Francisco J. Ayala, 211–19. San Diego : Elsevier/Academic Press.

Gauthey, B. 2025. Liberté d'expression : le dangereux retour du "saint-simonisme d'atmosphère" (Freedom of Expression: The Dangerous Return of 'Atmospheric Saint-Simonianism). *Le Point*, 15th January. https://www.lepoint.fr/debats/liberte-d-expression-le-dangereux-retour-du-saint-simonisme-d-atmosphere-15-01-2025-2580068_2.php.

Gaydarska, H., Takashima, K., Shahrier, S., and Raz, A. 2024. The interplay of ethics and genetic technologies in balancing the social valuation of the human genome in UNESCO declarations. *Eur. J. Human Genet.* 32: 725–730.

Geigl, E. M. 2020. Qui a habité en France ces 9 000 dernières années ? (Who has lived in France in the last 9,000 years?) *The Conversation*, 13th September. https://theconversation.com/qui-a-habite-en-france-ces-9-000-dernieres-annees-139769.

Gentillini, M., Caumes, E., Danis, M., Richard-Lenoble, D., Bégué, P. et al. 2012. *Médecine tropicale, 6° édition* (Tropical Medicine, 6th edition). Médecine Sciences publications, Lavoisier, Paris, France.

Gentillini, M., Duflo, B., Danis, M., Lagardère, B., Richard-Lenoble, D. et al.1986. Médecine tropicale, 4° édition. Médecine-Sciences, Flammarion, Paris, France.

Geschwind, D. H., and Flint, J. 2015. Genetics and genomics of psychiatric disease. *Science* 349: 1489–1494.

Gibbons, A. 2015. How Europeans evolved white skin. *Science*, 2nd April. https://www.science.org/content/article/how-europeans-evolved-white-skin.

Gibbons, A. 2017a. Neandertals mated early with modern humans. *Science* 356: 14.

Gibbons, A. 2017b. Oldest members of our species discovered in Morocco. *Science* 356: 993–94.

Gibbons, A. 2019. Spotting evolution among us. *Science* 363: 21–23.

Gibbons, A. 2025. Face to face with the first known Western European. *Science* 387: 1137.

Gichoya, J. W., Banerjee, I., Bhimireddy, A. R., Burns, J. L., Celi, L. A. et al. 2022. AI recognition of patient race in medical imaging: a modelling study. *Lancet Digit Health* 4(6): e406–e414. doi:10.1016/S2589-7500(22)00063-2.

Gilbert, E., Shanmugam, A., and Cavalleri, G. L. 2022. Revealing the recent demographic history of Europe via haplotype sharing in the UK Biobank. *Proc. Natl. Acad. Sci. USA* 119(25): e2119281119. https://doi.org/10.1073/pnas.2119281119.

Girardet, R. 1972. L'Idée coloniale en France de 1871 à 1962 (The Colonial Idea in France from 1871 to 1962). Paris, France : La Table Ronde.

Gobineau, A. 1853. Essai sur l'inégalité des Races humaines (Essay on the Inequality of the Human Races). Paris, France : Didot.

Gonzaga-Jauregui, C., Lupski, J. R., and Gibbs, R. A. 2012. Human Genome Sequencing in Health and Disease. *Annu. Rev. Med.* 63: 35–61.

González Burchard, E., Ziv, E., Coyle, N., Gomez, S. L., Tang, H. et al. 2003. The Importance of Race and Ethnic Background in Biomedical Research and Clinical Practice. *N. Engl. J. Med.* 348: 1170–1175.

Gould, S. J. 1996. The Mismeasure of Man. W. W. Norton & Company.

Gould, S. J. 2000. The spice of life. *Leader to leader* 15: 19–28.

Gourraud, P. A., Khankhanian, P., Cereb, N., Young Yang, S., Feolo, M. et al. 2014. HLA Diversity in the 1000 Genomes Dataset. *PLoS ONE* 9(7): e97282. doi:10.1371/journal.pone.0097282.

Gouveia, M. H., Meeks, K. A. C., Borda, V. Leal, T.P. Kehdy, F.S.G. et al. 2025. Subcontinental genetic variation in the All of Us Research Program: Implications for biomedical research. *Am. J. Human Genet.* 112: 1286–1301.

Grallet, G. 2018. Ces Chinois qui traquent les gènes de l'intelligence (Those Chinese who are hunting for the genes of intelligence). *Le Point*, 6th March. https://www.lepoint.fr/sciences-nature/ces-chinois-qui-traquent-les-genes-de-l-intelligence-06-03-2018-2199958_1924.php.

Grallet, G., and Mahler, T. 2019. Vers le retour d'une science raciste ? (Towards the return of a racist science?) *Le Point*, 13th September. https://www.lepoint.fr/societe/vers-le-retour-d-une-science-raciste-13-09-2019-2335481_23.php?M_BT=20588364922#xtor=EPR-6-[Newsletter-Matinale]-20190920.

Graves, J. L., and Goodman, A. H. 2022. *Racism, not race; answers to frequently asked questions.* Columbia university press.

Graves Jr., J. L., Goodman, A., Farina, S., and Moses, Y. 2025. US disregards scientific consensus on race *Science* 389: 242.

Guha, S., Rosenfeld, J. A., Malhotra, A. K., Lee, A T., Gregersen, P. K. et al. 2012. Implications for health and disease in the genetic signature of the Ashkenazi Jewish population. *Genome Biol.* 13: R2. doi : 10.1186/gb-2012-13-1-r2.

Günther, T., Valdiosera, C., Malmström, H., Ureña, I., Rodriguez-Varela, R. et al. 2015. Ancient genomes link early farmers from Atapuerca in Spain to modern-day Basques. *Proc. Natl. Acad. Sci. USA* 112 : 11917–11922.

Gunz, P., Tilot, A. K., Wittfeld, K., Teumer, A., Shapland, C.Y et al. 2019. Neandertal Introgression Sheds Light on Modern Human Endocranial Globularity. *Current Biol.* 29: 1–8. https://doi.org/10.1016/j.cub.2018.10.065.

Gurdasani, D., Carstensen, T., Tekola-Ayele, F., Pagani, L., Tachmazidou, I. et al. 2015. The African Genome Variation Project shapes medical genetics in Africa. *Nature* 517: 327–32.

Gurdasani, D., Barroso, I., Zeggini, E., and Sandhu, M. S. 2019. Genomics of disease risk in globally diverse populations. *Nature Rev. Genet.* 20: S20–S35.

Haak, W., Lazaridis, L., Patterson, N., Rohland, N., Mallick, S. et al. 2015. Massive migration from the steppe was a source for Indo-European languages in Europe. *Nature* 522: 207–11.

Haldane, J.B.S. 1949. Disease and Evolution. *La ricerca scientifica* Suppl 19: 68–76.

Hao, F., Zhu, W., Zheng, J., Chien Ng, S., and Zhang, J. 2025. Geographic diversity in inflammatory bowel disease genetics and microbiome. *Trends Microbiol.* 33: 1196–1211.

Harari, Y. N. 2011. *Sapiens. A Brief History of Humankind.* Harvill Secker, Random House, London, UK.

Hardelid, P., Cortina-Borja, M., Munro, A., Jones, H., Cleary, M. et al. 2008. The Birth Prevalence of PKU in Populations of European, South Asian and Sub-Saharan African Ancestry Living in South East England. *Ann. Human Genet.* 72: 65–71.

Harden, K. P. 2023. Genetic determinism, essentialism and reductionism: semantic clarity for contested science. *Nature Rev. Genet.* 24: 197–204.

Harpending, H. 2002. *Race. Population genetics perspective.* Encyclopedia of Evolution Tome II p. 979-981. Oxford University Press.

Harris, K. 2015. Evidence for recent, population-specific evolution of the human mutation rate. *Proc. Natl. Acad. Sci. USA* 112: 3439–3444.

Harris, K. M., Edlund, M. J., and Larson, S. 2005. Racial and ethnic differences in the mental health problems and use of mental health care. *Med. Care* 43: 775–784.

Hartl, D. L. 1985 *Our uncertain heritage: Genetics and human diversity (2nd ed.)* Harper & Row, New York.

Hebert, J. P. 1977. Race et intelligence (Race and Intelligence). Editions Copernic, Paris, France.

Henry, J. P. 2019. Génétique et origine *d'Homo sapiens* (Genetics and origin of *Homo sapiens*). *Médecine Sciences* 35: 39–45.

Herrnstein, R. J., and Murray, C. 1994. *The Bell Curve: Intelligence and Class Structure in American Life.* New York: The Free Press.

Herzig, A. F., Clerget-Darpoux F., and Génin E. 2022. The False Dawn of Polygenic Risk Scores for Human Disease Prediction. *J. Pers Med.* 266: doi: 10.3390/jpm12081266. PMID: 36013215; PMCID: PMC9409868.

Heuvelmans, B., and Porchnev, B. F. 1974. *L'Homme de Néanderthal est toujours vivant*, (Neanderthal man is still alive). Plon, Paris, France.

Hey, J. 2001. The mind of the species problem. *Trends Ecol. Evol.* 16: 326–329.

Heyer, E. 2025. L'odyssée d'"*Homo sapiens*", cet "animal curieux" qui migre depuis 300000 ans : conversation avec Evelyne Heyer. (The odyssey of "*Homo sapiens*," this "inquisitive animal" that has been migrating for 300,000 years: a conversation with Evelyne Heyer) *The Conversation*, 30th April. https://theconversation.com/lodyssee-d-homo-sapiens-cet-animal-curieux-qui-migre-depuis-300-000-ans-conversation-avec-evelyne-heyer-253925?utm_medium=email&utm_campaign=La%20newsletter%20du%20week-end%20-%203361834286&utm_content=La%20newsletter%20du%20week-end%20-%203361834286+CID_ce44f04db358d0f1908636d10bba100a&utm_source=campaign_monitor_fr&utm_term=Lodysse%20d%20Homo%20sapiens%20%20cet%20%20animal%20curieux.

Heyer, E., and Reynaud-Paligot, C. 2019. *On vient vraiment tous d'Afrique ?. Des préjugés au racisme : les réponses à vos questions* (Do we all really come from Africa? From prejudice to racism: answers to your questions). Editions Flammarion, Paris, France.

Highland, H. M., Wojcik, G. L., Graff, M., Young, K. L., Bien, S. A. et al. 2022. Predicted gene expression in ancestrally diverse populations leads to discovery of susceptibility loci for lifestyle and cardiometabolic traits. *Am. J. Human Genet.* 109: 669–679.

Hindorff, L. A., Bonham, V. L., Brody, L. C., Ginoza, M. E. C., Hutter, C. M. et al. 2018. Prioritizing diversity in human genomics research. *Nature Rev. Genet.* 19: 175–185.

Hochman, A. 2013. Racial discrimination: How not to do it. *Stud. Hist. Philos. Biol. Biomed. Sci.* 44: 278–286.

Hoffmann, T. J., Graff, R. E., Madduri, R. K., Rodriguez, A. A., Cario, C. L. et al. 2025. Genome-wide association study of prostate-specific antigen levels in 392,522 men identifies new loci and improves prediction across ancestry groups. *Nat Genet* 57: 334–344. https://doi.org/10.1038/s41588-024-02068-z.

Hofmann, V., Kalluri, P. R., Jurafsky, D., and King, S. 2024. AI generates covertly racist decisions about people based on their dialect. *Nature* https://doi.org/10.1038/s41586-024-07856-5.

Holloway, R. L. 2002. Head-to-head with Boas: Did he err on the plasticity of head form? *Proc. Natl. Acad. Sci. USA* 99: 14622–14623.

Hooton, E. A. 1926. Methods of racial analysis. *Science* 63: 75–81.

HoSang, D. M. 2014. On Racial Speculation and Racial Science: A Response to Shiao et al. *Sociological Theory* 32: 228–43.

Hsu, C., Yang, W., Parikh, R.V., Anderson, A. H., Chen, T. K. et al. 2021. Race, Genetic Ancestry, and Estimating Kidney Function in CKD. *N. Engl. J. Med.* 385: 1750–1760.

Hublin, J. J. 2011. *Quand d'autres hommes peuplaient la terre* (When other men populated the earth). Paris, France : Flammarion.

Hublin, J. J., Ben-Ncer, A., Bailey, S. E., Freidline, S. E., Neubauer, S. et al. 2017. New fossils from Jebel Irhoud, Morocco and the pan-African origin of *Homo sapiens*. *Nature* 546: 289–305.

Hublin, J. J., Sirakov, N., Aldeias, V., Bailey, S., Bard, E. et al. 2020. Initial Upper Palaeolithic *Homo sapiens* from Bacho Kiro Cave, Bulgaria. *Nature* 581: 299–318.

Huffman, J. E., Gaziano, L., Zeina, R., Al Sayed, Z. R., Renae, L. et al. 2025. An African ancestry-specific nonsense variant in CD36 is associated with a higher risk of dilated cardiomyopathy. *Nature Genet.* 57: 2682–2690.

Huguet, R., Rodríguez-Álvarez, X. P., María Martinón-Torres, M., Vallverdú, J., López-García, J.M. et al. 2025. The earliest human face of Western Europe. *Nature* 640: 707–713.

Huston, N., and Raymond, M. 2013. Oui, les races existent ! (Yes, races exist !) *Le Monde*, 17th May.

Huxley, J. S. 1924. America Revisited, 3: The Negro Problem. *The Spectator*, 29 November : 821.

Hyman, S. 2014. Depression needs large human-genetics studies. *Nature* 515: 189–191.

Iasi, L. N. M., Chintalapati, M., Skov, L., Mesa, A. B., Hajdinjak, M. et al. 2024. Neanderthal ancestry through time : Insights from genomes of ancient and present-day humans. *Science* 386: 1239

Jacquard, A.1978. Eloge de la différence: la génétique et les hommes (In Praise of Difference: Genetics and Men). Editions du Seuil Paris, France.

Jacquard, A. 1982. Au péril de la science ? Interrogations d'un généticien (At the risk of science? Questions from a geneticist). Editions du Seuil, Paris, France.

Jaratlerdsiri, W., Jiang, J., Gong, T., Patrick, S. M., Willet, C. et al. 2022. African-specific molecular taxonomy of prostate cancer. *Nature* 609: 552–559.

Ji, Y., Liu, N., Yang, Y., Wang, M., Cheng, J. et al. 2025. Cross-ancestry and sex-stratified genome-wide association analyses of amygdala and subnucleus volumes. *Nature Genet.* 57: 839–850.

Jia, G., Chen, Z., Ping, J., Cai, Q., Tao, R. et al. 2025. Refining breast cancer genetic risk and biology through multi-ancestry fine-mapping analyses of 192 risk regions. *Nature Genet.* 57: 80–87

Johnston, I. D. A., Bland, J. M., and Anderson, H. R. 1987. Ethnic variation in respiratory morbidity and lung function in childhood. *Thorax* 42: 542–548.

Jonsson, H., Magnusdottir, E., Eggertsson, H. P., Stefansson, O. A., Arnadottir, G. A. et al. 2021. Differences between germline genomes of monozygotic twins. *Nature Genet.* 53: 27–34.

Jordan, B. 2008. L'humanité au pluriel. La génétique et la question des races (Mankind in the plural Genetics and the question of races). Paris, France : Éditions du Seuil.

Jordan, B. 2010. ADN, ascendance génétique et "race" sociale : l'apport des Snip (DNA, genetic ancestry and social "race": the contribution of Snips). *Médecine/Sciences* 26 : 215–218.

Jordan, B. 2021. La génomique et la diversité humaine (genomics and human diversity). *Cahiers de l'Urmis* http://journals.openedition.org/urmis/2387.

Jordan, B. 2025 Séquençage de l'ADN. La fin d'un quasi-monopole (DNA sequencing. The end of a quasi-monopoly). *Médecine/Sciences* 41: 611–614.

Jorde, L. B., and Wooding, S. P. 2004. Genetic variation, classification and 'race'. *Nature Genet.* suppl. 11 : S28–S33.

Joshi, P. K., Esko, T., Mattsson, H., Eklund, N., Gandin, I. et al. 2015. Directional dominance on stature and cognition in diverse human populations. *Nature* 523: 459–62.

Joubin, L., and Robin, A. 1923. *Les animaux, les invertébrés et les vertébrés. Histoire Naturelle* (Animals, invertebrates and vertebrates. Natural History). Paris, France : Larousse.

Ju, D., Hui, D., Dorothy A., Hammond, D. A., Wonkam, A. et al. 2022. Importance of Including Non-European Populations in Large Human Genetic Studies to Enhance Precision Medicine. *Annu. Rev. Biomed. Data Sci.* 5: 321–339.

Kaiser, J. 2023. Panel urges caution in tying social, behavioral traits to genes. *Science* 380: 327–328.

Kaplan, J. M. 2015. Race, IQ, and the search for statistical signals associated with so-called "X"-factors: environments, racism, and the "hereditarian hypothesis". *Biol. Philos.* 30: 1–17; DOI 10.1007/s10539-014-9428-0.

Karczewski, K. J., Gupta, R., Kanai, M., Lu, W., Tsuo, K. et al. 2025. Pan-UK Biobank genome-wide association analyses enhance discovery and resolution of ancestry-enriched effects. *Nature Genet.* 57: 2408–2417.

Karlsson, E. K., Kwiatkowski D. P., and Sabeti, P. C. 2014. Natural selection and infectious disease in human populations. *Nature Rev. Genet.* 15: 379–83.

Khan, A.T., Gogarten, S. M., McHugh, C. P., Stilp, A. M., Sofer, T. et al. 2022. Recommendations on the use and reporting of race, ethnicity, and ancestry in genetic research: Experiences from the NHLBI TOPMed program. *Cell Genomics* https://doi.org/10.1016/j.xgen.2022.100155.

Kierkegaard, E.O.W. 2019. Race Differences: A Very Brief Review. *The Mankind Quarterly*, December. DOI: 10.46469/mq.2019.60.2.2.

Kim, H. J., Yang, S. G., Greenspan, J. D., Downton, K., Griffith, K. A. et al. 2017. Racial and ethnic differences in experimental pain sensitivity: systematic review and meta-analysis. *Pain* 158: 194–211. DOI: 10.1097/j.pain.0000000000000731.

King, M. C., and Wilson, A. C. 1975. Evolution at two levels in humans and chimpanzees. *Science* 188: 107–116.

King, R. C., and Stanfield, W. D. 1990. A dictionary of genetics 4th edition. Oxford University Press, New York, Oxford.

King, R.C., Mulligan, P. K., and Stanfield, W. D. 2012. *A dictionary of genetics* 8th edition. Oxford University Press, New York, Oxford.

Klarsfeld, A., and Vernier, P. 1998. La biologie au service de l'antiracisme : limites et dangers (Biology in the service of anti-racism: limits and dangers). *Médecine/Sciences* 11: 1292–1294.

Knopik, V.S. 2015. *Behavioral Genetics : Psychological Perspectives*. In *International Encyclopedia of the Social & Behavioral Sciences*, 2nd edition, Volume 2, edited by James Wright, 450–55. Elsevier Ltd. https://doi.org/10.1016/B978-0-08-097086-8.82035-8.

Koch, L. 2022. Impact of genetic ancestry on viral infection response. *Nat. Rev. Genet.* 23: 72. https://doi.org/10.1038/s41576-021-00442-9.

Kodjo-Grandvaux, S. 2021. Ceux qui disent que tout est "race" ont autant tort que ceux qui disent que rien n'est "race"(Those who say that everything is "race" are as wrong as those who say that nothing is "race."). *Le Monde*, 23rd November. https://www.lemonde.fr/idees/article/2021/11/23/catherine-coquery-vidrovitch-ceux-qui-disent-que-tout-est-race-ont-autant-tort-que-ceux-qui-disent-que-rien-n-est-race_6103201_3232.html

Kodjo-Grandvaux, S. 2025. Michel Agier, anthropologue : La race biologique n'existe pas, mais la race produite par le racisme, elle, existe socialement. (Biological race does not exist, but race produced by racism does exist socially). *Le Monde*, 10th February. https://www.lemonde.fr/idees/article/2025/02/10/michel-agier-anthropologue-la-race-biologique-n-existe-pas-mais-la-race-produite-par-le-racisme-elle-existe-socialement_6539488_3232.html.

Korzeniewski K., Bylicka-Szczepanowska, E., and Lass, A. 2021. Prevalence of Asymptomatic Malaria Infections in Seemingly Healthy Children, the Rural Dzanga Sangha Region, Central African Republic. *Int. J. Environ. Res. Public Health* 18: 814. https://doi.org/10.3390/ ijerph18020814.

Krause, J., Lalueza-Fox, C., Orlando, L., Enard, W., Green, R. E. et al. 2007. The Derived FOXP2 Variant of Modern Humans Was Shared with Neandertals. *Current Biol.* 17: 1908–12.

Kronfeldner, M. 2017. The politics of human nature. pp. 625–32. *In:* Michel Tibayrenc and Francisco J. Ayala (eds.). *On Human Nature: Biology, Psychology, Ethics, Politics, and Religion*. San Diego: Elsevier/Academic Press.

Kullo, I. J. 2024. Promoting equity in polygenic risk assessment through global collaboration. *Nature Genet.* 56: 1780–1787.

Kumar, R., Seibold, M. A., Aldrich, M. C., Williams, L. K., M. P. H., Reiner, A. et al. 2010. Genetic Ancestry in Lung-Function Predictions. *N. Engl. J. Med.* 363: 321–330.

Kuhlwilm, M., Gronau, I., Hubisz, M., de Filippo, C., Prado-Martinez, J. et al. 2016. Ancient gene flow from early modern humans into Eastern Neanderthals. *Nature* 530: 429–33.

Lachance, J., and Tishkoff, S. A. 2013. Population Genomics of Human Adaptation. *Annu. Rev. Ecol. Evol. Syst.* 44: 123–143.

Lahn, B. T., and Ebenstein, L. 2009. Let's celebrate human genetic diversity. *Nature* 461: 726–728.

Lala, K. L., and Feldman, M. W. 2024. Genes, culture, and scientific racism. *Proc. Natl. Acad. Sci. USA* 121: e2322874121.

Laland, K. N., Odling-Smee, J., and Myles, S. 2010. How culture shaped the human genome: bringing genetics and the human sciences together. *Nature Rev. Genet.* 11: 137–48.

Lamoureux, N. 2025. Une entreprise américaine a-t-elle vraiment ressuscité le "loup terrible" des âges glaciaires ? (Has an American company really resurrected the 'dire wolf' of the Ice Ages?). *Le Point*, 9th April. https://www.lepoint.fr/science/comment-une-entreprise-americaine-ressuscite-le-loup-terrible-des-ages-glaciaires-09-04-2025-2586951_25.php.

Lander, E. S. 2011. Initial impact of the sequencing of the human genome. *Nature* 470: 187–197.

Langaney, A. 1999. *La philosophie biologique (Regards sur la science)* (Biological philosophy (Perspectives on science)). Paris, France : Belin.

Lasker, J., Pesta, B. J., Fuerst, J. G. R., and Kirk, E. O. 2019. Global Ancestry and Cognitive Ability. *Psych* 1: 431–459.

Last, C. 2012. The 10,000 Year Explosion: How Civilization Accelerated Human Evolution (2009) by Gregory Cochran and Henry Harpending. vis-à-vis: Explorations in Anthropology, Vol.XX, No. X, pp. XX–XX.

Lawler, J. M. 1978. *IQ, heritability, and racism*. New York : International Publishers.

Lazaridis, I., Patterson, N., Anthony, D., Vyazov, L., Fournier, R. et al. 2025. The genetic origin of the Indo-Europeans. *Nature*: https://doi.org/10.1038/s41586-024-08531-5.

Lazer, D. M. J., Baum, M. A., Benkler, Y., Berinsky, A. J., Greenhill, K. M. et al. 2018. The science of fake news. *Science* 359: 1094–96.

Le Bars, S. 2021. La "critical race theory", nouvel avatar de la guerre culturelle aux États-Unis ("Critical race theory," a new manifestation of the cultural war in the United States). *Le Monde,* 2nd July. https://www.lemonde.fr/idees/article/2021/07/02/la-crt-ou-critical-race-theory-nouvel-avatar-de-la-guerre-culturelle-aux-etats-unis_6086634_3232.html.

Lee, J. 2022. Don't Even Go There. *City Journal*, October 19. https://www.city-journal.org/article/dont-even-go-there.

Lee, J. J., Wedow, R., Okbay, A., Kong, E., Maghzian, O. et al. 2018. Gene discovery and polygenic prediction from a genome-wide association study of educational attainment in 1.1 million individuals. *Nature Genet.* 50: 112–1121.

Lee, S. S. J., and Mudaliar, A. 2009. Racing forwards: the Genomics and Personalized Medicine Act. *Science* 323: 342.

Leroi, A. M. 2005. A Family Tree in Every Gene. *The New York Times,* March 14.

Leroyer, A. 2024. Le passage des chasseurs-cueilleurs à l'agriculture a été beaucoup plus violent qu'on ne l'imaginait (The transition from hunter-gatherers to agriculture was far more violent than previously thought). *Huffington Post*, 20th February. https://www.huffingtonpost.fr/science/video/le-passage-des-chasseurs-cueilleurs-a-l-agriculture-a-ete-beaucoup-plus-violent-qu-on-ne-l-imaginait_230013 html.

Leslie, S., Winney, B., Hellenthal, G., Davison, D., Boumertit, A. et al. 2015. The fine-scale genetic structure of the British population. *Nature* 519: 309–14.

Lévi-Strauss, C. 1952. Race et histoire (Race and History). UNESCO Editions.

Lévi-Strauss, C. 1971. Race et culture (Race and Culture). UNESCO editions.

Lewino, F. 2020. Sapiens : on a retrouvé le plus vieil Européen… en Bulgarie! (Sapiens: The oldest European has been found… in Bulgaria!) *Le Point*, 11th May. https://www.lepoint.fr/sciences-nature/sapiens-on-a-retrouve-le-plus-vieil-europeen-en-bulgarie-11-05-2020-2375042_1924.php.

Lewis, A. C. F., Molina, S. J., Appelbaum, P. S., Dauda, B., Di Rienzo, A. D. et al. 2022. Getting genetic ancestry right for science and society. *Science* 376: 250–252.

Lewis, J. E., DeGusta, D., Meyer, M. R., Monge, J. M., Mann, A. E. et al. 2011. The Mismeasure of Science: Stephen Jay Gould versus Samuel George Morton on Skulls and Bias. *PLoS Biol.* 9(6): e1001071. https://doi.org/10.1371/journal.pbio.1001071.

Lewontin, R. C. 1972. The apportionment of human diversity. *Evol. Biol.* 6: 381–98.

Lewontin, R. C., Rose, S., and Kamin, L. J. 1984. *Not in Our Genes: Biology, Ideology, and Human Nature.* New York: Pantheon Books.

Lherminier, P., and Solignac, M. 2005. *De l'espèce. Chapitre 13. Pluralisme de l'espèce.* (About species. Chapter 13. Pluralism of the species) Editions Syllepse, Paris, France.

Li, Y., Giorgi, E. E., Beckman, K. B., Caberto, C., Kazma, R. et al. 2019. Association between mitochondrial genetic variation and breast cancer risk: The Multiethnic Cohort. *PLoS ONE* 14(10): e0222284. https://doi.org/10.1371/journal.pone.0222284.

Lintzeri, D. A., Karimian, N., Blume-Peytavi, U., and Kottner, J. 2022. Epidermal thickness in healthy humans: a systematic review and meta-analysis. *JEADV* 36: 1191–1200.

Liu, N., and Liu, C. 2023. Cross-ancestry genetic discovery for hippocampal volumetric traits. *Nature Genet.* 55: 1086–1087.

Livingstone F. 1962. On the non-existence of human races. *Curr. Anthropol.* 3: 279–281.

López Herráez, D., Bauchet, M., Tang, K., Theunert, C., IPugach, I., Li, J. et al. 2009. Genetic variation and recent positive selection in worldwide human populations: evidence from nearly 1 Million SNPs. *Plos One* 4 (11): e7888. doi:10.1371/journal.pone.0007888

Lordkipanidze, D. 2017. The History of Early *Homo.* In *On Human Nature: Biology, Psychology, Ethics, Politics, and Religion*, edited by Michel Tibayrenc and Francisco J. Ayala, 45–54. San Diego: Elsevier/Academic Press.

Louvet, B. 2024a. Combien de temps ont duré les croisements entre Néandertaliens et humains modernes? (How long did interbreeding between Neanderthals and modern humans last?) *SciencePost*, 13th December. https://sciencepost.fr/combien-de-temps-croisements-neandertaliens-et-humains-modernes/.

Louvet, B. 2024b. Les plus anciennes gravures "intentionnelles" d'Europe ne sont pas l'œuvre *d'Homo Sapiens* (The oldest "intentional" engravings in Europe are not the work of *Homo sapiens*). *SciencePost*, 15th December. https://sciencepost.fr/les-plus-anciennes-gravures-ne-sont-pas-dhomo-sapiens/.

Lynn, R., and Vanhanen, T. 2002. IQ and the wealth of nations. Greenwood Publishing Group.

Maglo, K. N., Mersha, T. B., and Martin, L. J. 2016. Population Genomics and the Statistical Values of Race: An Interdisciplinary Perspective on the Biological Classification of Human Populations and Implications for Clinical Genetic Epidemiological Research. *Front. Genet.* 7: 22. doi: 10.3389/fgene.2016.00022.

Malik, K. 2012. Why both sides are wrong in the race debate. *Pandaemonium.* https://kenanmalik.wordpress.com/2012/03/04/why-both-sides-are-wrong-in-the-race-debate/.

Mallapaty, S. 2025. Trump team 'survey' sent to overseas researchers prompts foreign interference fears. *Nature*, News, 20 March. *doi:* https://doi.org/10.1038/d41586-025-00850-5.

Mangin, L. 2023. La péninsule arabique, un incubateur pour… *Homo sapiens* (The Arabian Peninsula, an incubator for… *Homo sapiens*). *Pour la Science.* https://www.pourlascience.fr/sd/paleontologie-humaine/la-peninsule-arabique-un-incubateur-pour-homo-sapiens-25467.php?utm_source=email&utm_medium=email&utm_campaign=nl_pls_hebdo_153

Manry, J., and Quintana-Murci, L. 2013. A Genome-Wide Perspective of Human Diversity and Its Implications in Infectious Disease. *Cold Spring Harb. Perspect. Med.* 2013;3:a012450.

Marouli, E., Graff, M., Medina-Gomez, C., Ken Sin Lo, K. S., Wood, A. R. et al. 2017. Rare and low-frequency coding variants alter human adult height. *Nature* 542: 186–193.

Marshall, M. 2025. Women's pelvises are shrinking – how is that changing childbirth? *New Scientist*, 24th June. https://www.newscientist.com/article/2485371-womens-pelvises-are-shrinking-how-is-that-changing-childbirth/.

Martin, A. R., Gignoux, C. R., Walters, R. K., Wojcik, G. L., Neale, B. M. et al. 2017. Human Demographic History Impacts Genetic Risk Prediction across Diverse Populations. *Am. J. Human Genet.* 100: 635–649.

Martini, R., and Davis, M. B. 2024. The DARC side of genetics in cancer: Breast cancer disparities. *Am. J. Human Genet.* 111: 1261–1264.

Martinón-Torres, M., and Lalueza-Fox, C. 2025. Clues about the earliest migrations out of Africa. *Nature* 638: 620–621.

Mary, C. 2015. La génomique réécrit l'histoire (Genetics Rewrites History). *Le Monde*, 16th June. https://www.lemonde.fr/sciences/article/2015/06/08/la-genomique-reecrit-l-histoire_4649802_1650684.html.

Mary, C. 2018a. 'Race' : la génétique face à ses démons. (Race: Genetics Facing Its Demons) *Le Monde*, 11th July. https://abonnes.lemonde.fr/sciences/article/2018/07/10/race-la-genetique-face-a-ses-demons_5328964_1650684.html.

Mary, C. 2018b. Il y a un fort retour de la thématique de la race du côté de la génomique et de la biomédecine (There is a strong return of the theme of race in genomics and biomedicine.). *Le Monde*, 10th July. https://www.lemonde.fr/sciences/article/2018/07/10/il-y-a-un-fort-retour-de-la-the-matique-de-la-race-du-cote-de-la-genomique-et-de-la-biomedecine_5328965_1650684.html.

Mathieson, I., Lazaridis, I., Rohland, N., Mallick, S., Patterson, N. et al. 2015. Eight thousand years of natural selection in Europe. Preprint at bioRxiv. https://www.biorxiv.org/content/10.1101/016477v2

Mauro, M., Allen, D. S., Dauda, B., Molina, S. J., Neale, B. M. et al. 2022. A scoping review of guidelines for the use of race, ethnicity, and ancestry reveals widespread consensus but also points of ongoing disagreement. *Am. J. Human. Genet.* 109: 2110–2125.

Mayr, E. 1963. *Animal species and evolution.* Cambridge, MA : Belknap Press of Harvard University Press.

Mayr, E. 2002. The Biology of Race and the Concept of Equality. *Daedalus*: 131: 89–94.

Mayr, E., Linsley, E. G., and Usinger, R. L. 1953. Methods and principles of systematic zoology. McGraw-Hill, New York.

McLaren, P. J., Porreca, I., Iaconis, G., Mok, H. P., Mukhopadhyay, S. et al. 2023. Africa-specific human genetic variation near CHD1L associates with HIV-1 load. *Nature* 620: 1025–1030.

Meloni, M. 2017. Race in an epigenetic time: thinking biology in the plural. *British J. Sociol.* 68: DOI: 10.1111/1468-4446.12248.

Mersha, T. B., and Abebe, T. 2015. Self-reported race/ethnicity in the age of genomic research: its potential impact on understanding health disparities. *Hum. Genomics*. 9: 1–15. DOI 10.1186/s40246-014-0023-x.

Mervis, J. 2022. NSF grant decisions reflect systemic racism, study argues. *Science* 377: 455–456.

Miao, J. 2025. Understanding nature and nurture. *Science* 390: 584.

Michel, J.B. 2018. Préhistoire : il y a 2 millions d'années, le long voyage des premiers hommes. (Prehistory: 2 million years ago, the long journey of the first men.) *GÉO Histoire* 25th July. https://www.geo.fr/photos/reportages-geo/le-long-voyage-des-premiers-hommes-188016.

Mohsen, H. 2020. Race and Genetics: Somber History, Troubled Present. *Yale J. Biol. Med.* 93: 215–218.

Molteni, M. 2022. Buffalo shooting ignites a debate over the role of genetics researchers in white supremacist ideology. *STATnews*, 23 mai. https://www.statnews.com/2022/05/23/buffalo-shooting-ignites-debate-genetics-researchers-in-white-supremacist-ideology/.

Moreno-Estrada, A., Gignoux,, C. R., Fernández-López, J. C., Zakharia, F., Sikora, M. et al. 2014. The genetics of Mexico recapitulates Native American substructure and affects biomedical traits. *Science* 344: 1280–85.

Mounk, Y. 2024. Nazis noirs, femmes papes… La leçon woke de l'IA Google Gemini (Black Nazis, female popes… The woke lesson of Google Gemini AI). *Le Point*, 2nd March. https://www.lepoint.fr/postillon/nazis-noirs-femmes-papes-google-gemini-l-ia-maitresse-d-ecole-woke-02-03-2024-2554044_3961.php?boc=208785&nl_key=1a4dad71824665c5a198b17610c4f0a226f370549e46babfbc4da61c2f46680b&mi=jby9fp0KZ7CvY3KJnLC8WK1lQ04yMN_DKejaXLltVi%2BSPplDrL82L8QyVGsCNEC9%2BQDDwx9CFwJguU4mhfn%2BvEP9ECyjjD&M_BT=20588364922#xtor=EPR-57-[Push-email]-20240302-[Article_1].

Morin, H. 2024. Médecine prédictive : sur les scores polygéniques, les limites du tout-génétique (Predictive medicine: on polygenic scores, the limits of all-genetics.). *Le Monde*, 11th December. https://www.lemonde.fr/sciences/article/2024/12/11/medecine-predictive-sur-les-scores-polygeniques-les-limites-du-tout-genetique_6441530_1650684.html.

Morin, H. 2025. L'inquiétant retour de l'eugénisme génétique. (The worrying return of genetic eugenics) *Le Monde*, 14th January. https://www.lemonde.fr/idees/article/2025/01/14/l-inquietant-retour-de-l-eugenisme-genetique_6497151_3232.html.

Morton, S. G. 1849. Catalogue of skulls of man and the inferior animals, Third Edition. Philadelphia: Merrihew and Thomson Printers.

Mountain, J. L., and Risch, N. 2004. Assessing genetic contribution to phenotypic differences among "racial" and "ethnic" groups. *Nature Genet. Suppl.* 36: S48–S53.

Murray, C. 2020. *Human diversity. The biology of gender, race, and class*. TWELVE New York, Boston.

Narasimhan, V. M., Patterson, N., Moorjani, P., Rohland, N., Bernardos, R. et al. 2019. The formation of human populations in South and Central Asia. *Science* 365, eaat7487 (2019). DOI: 10.1126/science.aat7487.

National Academies of Sciences, Engineering, and Medicine. 2024. *Rethinking Race and Ethnicity in Biomedical Research.* Washington, DC: The National Academies Press. https://doi.org/10.17226/27913 .

Need, A. C., Kasperavičiūtė, D., Cirulli, E. T., and Goldstein, D. B. 2009. A genome-wide genetic signature of Jewish ancestry perfectly separates individuals with and without full Jewish ancestry in a large random sample of European Americans. *Genome Biol.* 10: R7.

Neel, J. V. 1962. Diabetes Mellitus: A 'Thrifty' Genotype Rendered Detrimental by 'Progress'? *Am. J. Human Genet.* 14: 353–62.

Nei, M. 1978. The theory of genetic distance and evolution of human races. *Japanese J. Human Genet.* 23: 341–69.

Nei, M., and Takezaki, N. 1996. The Root of the Phylogenetic Tree of Human Populations. *Mol. Biol. Evol.* 13: 170–177.

Nelson, R. G. 2019. Racism in science: the taint that lingers. *Nature* 570: 440–44

Nelson, M. R., Wegmann, D., Ehm,, M. G., Kessner, D., Jean, P. S. et al. 2012. An Abundance of Rare Functional Variants in 202 Drug Target Genes Sequenced in 14,002 People. *Science* 337: 100–104.

News at a glance 2022. Pediatricians take stand on race. *Science* 376: 677

News at a glance. 2023. Human geneticist apologize. *Science* 379: 314.

Niedzwiedz, C. L., O'Donnell, C. A., Jani, B. D., Demou, E., Ho, F. K. et al. 2020. Ethnic and socioeconomic differences in SARS-CoV-2 infection: prospective cohort study using UK Biobank. *BMC Medicine* 18: 160.

Ninio, J. 1979. Approches moléculaires de l'évolution (Molecular approaches to evolution). Masson, éd. Paris, France.

Nobles, M., Womack, C., Wonkam A., and Wathuti, E. 2022. Ending racism is key to better science: a message from *Nature*'s guest editors. *Nature* 610: 419–420.

Novembre, J. 2015. Ancient DNA steps into the language debate. *Nature* 522: 164–65.

Novembre, J., and Han, E. 2012. Human population structure and the adaptive response to pathogen-induced selection pressures. *Phil. Trans. R. Soc. B* 367: 878–886.

Novembre, J., Johnson, T., Bryc, K., Kutalik, Z., Boyko, A. R. et al. 2008. Genes mirror geography within Europe. *Nature* 456: 98–102.

Okbay, A., Wu, Y., Wang, N., Jayashankar, H., Bennett, M. et al. 2022. Polygenic prediction of educational attainment within and between families from genome-wide association analyses in 3 million individuals. *Nature Genet.* 54: 437–449.

Omiye, J. A., Lester, J. C., Spichak, S., Rotemberg, V., and Daneshjou, R. 2023. Large language models propagate race-based medicine. *npj Digit. Med.* 6: 195. https://doi.org/10.1038/s41746-023-00939-z

Oni-Orisan, A., Mavura, Y., Banda, Y., Thornton, T. A., and Sebro, R. 2021. Embracing genetic diversity to improve black health. *New Engl. J. Med.* 384: 1163–1167.

Orlando, L. 2021. Ce biais cognitif qui nous divise (this cognitive bias that divides us). *Le Point Phébé*, 30th November. https://www.lepoint.fr/phebe/phebe-ce-biais-cognitif-qui-nous-divise-30-11-2021-2454440_3590.php?M_BT=20588364922#xtor=EPR-6-[]-20211204.

Ostrer, H., and Skorecki, K. 2013. The population genetics of the Jewish people. *Human Genet.* 132: 119–127.

Ousley, S., Jantz, R., and Freid, D. 2009. Understanding Race and Human Variation: Why Forensic Anthropologists are Good at Identifying Race. *Amer. J. Physical Anthropol.* 139: 68–76.

Oza, A. 2024. Why is heart disease more common in people from South Asia? *Nature* 633: 272–274.

Pääbo, S. 2015. The diverse origins of the human gene pool. *Nature Rev. Genet.* 16: 313–14.

Pääbo, S. 2016. Je cherche à identifier ce qui rend les hommes modernes uniques. (I seek to identify what makes modern humans unique) *Le Temps*, 8th November. https://www.letemps.ch/sciences/cherche-identifier-rend-hommes-modernes-uniques.

Pankratov, V., Mezzavilla, M., Aneli, S., Kuznetsov, I. A., Fusco, D. et al. 2024. Ancestral genetic components are consistently associated with the complex trait landscape in European biobanks. *Eur. J. Human Genet.* 32: 1492–1499.

Parens, E. 2020. The Inflated Promise of Genomic Medicine. *Sci. Am.* 6th January. https://blogs.scientificamerican.com/observations/the-inflated-promise-of-genomic-medicine/.

Parra, E. J., Kittles, R. A., and Shriver, M. D. 2004. Implications of correlation between skin color and genetic ancestry for biomedical research. *Nature Genet. Suppl.* 36: S54–S60.

Patou-Mathis, M. 2013. De la hiérarchisation des êtres humains au paradigme racial (From the hierarchy of human beings to the "racial paradigm"). In Hermès, La Revue 2013/2 (n° 66), pages 30–37, CNRS Ed.

Patrinos, A. 2004. Race and the human genome. Sponsor's forward. *Nature Genet. Suppl.* 36: S1–S2.

Pearson, H. 2025. What's really driving the rise of autism? *Nature* 644: 860–863.

Peifer, M. 2025. We must not ignore eugenics in our genetics curriculum. *Trends Genet.* 41: 261–265.

Pesta, B. J., Kirkegaard, E. O. W., te Nijenhuis, J., Lasker, J., and Fuersta, J. G. R. 2020. Racial and ethnic group differences in the heritability of intelligence: A systematic review and meta-analysis. *Intelligence* 78: 101408.

Peterson, R. E., Kuchenbaecker, K., Walters, R. K., Chen, C. Y., Popejoy, A. B. et al. 2019. Genome-wide Association Studies in Ancestrally Diverse Populations: Opportunities, Methods, Pitfalls, and Recommendations. *Cell* 179: https://doi.org/10.1016/j.cell.2019.08.051.

Pichot, A. 1996. Racisme et biologie (Racism and Biology). *Le Monde*, 4th October. https://www.lemonde.fr/archives/article/1996/10/04/racisme-et-biologie_3746286_1819218.html.

Pinker, S. 2003. The Blank Slate: The Modern Denial of Human Nature. London : Penguin Books.

Pinker, S. 2004. Why nature & nurture won't go away. *Daedalus* 133: 5–17.

Plomin, R., and von Stumm, S. 2018. The new genetics of intelligence. *Nature Rev. Genet.* 191: 148–158.

Pontarotti, G. 2021. Il paraît souhaitable de ne pas balayer d'un revers de la main toute discussion sur la question dite "raciale" (It seems desirable not to dismiss out of hand all discussion on the so-called "racial" question). *Le Monde*, 14th May. https://www.lemonde.fr/idees/article/2021/05/14/il-parait-souhaitable-de-ne-pas-balayer-d-un-revers-de-la-main-toute-discussion-sur-la-question-dite-raciale_6080234_3232.html.

Pontarotti, G. 2023. Race et biologie à l'ère de l'épigénétique. Naturalisme, environnementalisme, constructivisme (Race and Biology in the Age of Epigenetics: Naturalism, Environmentalism, Constructivism). *Dialogue* 62: 279–301.

Popejoy, A. B., Ritter, D. I., Crooks, K., Currey, E., Fullerton, S. M. et al. 2018. The clinical imperative for inclusivity: Race, ethnicity, and ancestry (REA) in genomics. *Hum. Mutat.* 39: 1713–1720.

Popejoy, A. B. 2021. Too many scientists still say Caucasian. *Nature* 596: 463.

Popper, K. 1973. Logique de la découverte scientifique (Logic of scientific discovery). Payot éd., Paris, France.

Press Release 2020. New AMA policies recognize race as a social, not biological, construct. AMA Media & Editorial. https://www.ama-assn.org/press-center/press-releases/new-ama-policies-recognize-race-social-not-biological-construct.

Press Releases 2023. ATS Publishes Official Statement on Race, Ethnicity and Pulmonary Function Test Interpretation. ATS. American Thoracic Society https://www.thoracic.org/about/newsroom/press-releases/journal/2023/pft-and-race-official-statement.php.

Privé, F., Aschard, H., Carmi, S., Folkersen, L., Hoggart, C. et al. 2022. Portability of 245 polygenic scores when derived from the UK Biobank and applied to 9 ancestry groups from the same cohort *Am. J. Hum. Genet.* 109: 12–23.

Quintana-Murci, L. 2021. *Le peuple des humains* (The People of Humans). Editions Odile Jacob, Paris, France.

Racimo, F., Sankararaman, S., Nielsen, R., and Huerta-Sánchez, E. 2015. Evidence for archaic adaptive introgression in humans. *Nature Rev. Genet.* 16: 359–71.

Radier, V. 2020. Il y a 40 000 ans, plusieurs espèces humaines peuplaient la Terre (un peu comme dans Tolkien) (40,000 years ago, several human species populated the Earth (a bit like in Tolkien). *L'Obs*, 23rd February. https://www.nouvelobs.com/idees/20200223.OBS25182/il-y-a-40-000-ans-plusieurs-especes-humaines-peuplaient-la-terre-un-peu-comme-dans-tolkien.html#modal-msg.

Raff, J. 2014. Nicholas Wade and Race: Building a Scientific Façade, *Human Biol.* 86: http://digitalcommons.wayne.edu/humbiol/vol86/iss3/7.

Rambert, H. 2022. Arrière-grand-père fumeur, le poids en héritage (Great-grandfather smoker, weight as a legacy). *Le Point*, 31st January. https://www.lepoint.fr/sante/arriere-grand-pere-fumeur-le-poids-en-heritage-31-01-2022-2462801_40.php?M_BT=20588364922#xtor=EPR-6-[Newsletter-Matinale]-20220201-[Article_9.

Rao, V. M., Hla, M., Moor, M., Adithan, S., Kwak, S. et al. 2025. Multimodal generative AI for medical image interpretation. *Nature* 639: 888–896.

Reardon, S. 2023. Alzheimer's drug trials plagued by lack of racial diversity. *Nature* 620: 256–257.

Reich, D. 2018a. How Genetics Is Changing Our Understanding of 'Race'. *New York Times*, March 23rd.

Reich, D. 2018b. *Who we are an how we got there*. New York: Pantheon books.

Reich, D. 2018c. How to Talk About 'Race' and Genetics. *New York Times*, March 30th.

Reid, D. M., Barber, R. C., Thorpe Jr., R. J., Sun, J., Zhou, Z. et al. 2022. Mitochondrial DNA oxidative mutations are elevated in Mexican American women potentially implicating Alzheimer's disease. *npj Aging*: 8:2. https://doi.org/10.1038/s41514-022-00082-1.

Relethford, J. H. 2009. Race and Global Patterns of Phenotypic Variation. *Amer. J. Physic. Anthropol.* 139: 16–22.

Richardson, S. S. 2011. Race and IQ in the postgenomic age: The microcephaly case. *BioSocieties* 6: 420–46.

Richerson, P. J., Boyd, R., and Henrich, J. 2010. Gene-culture coevolution in the age of genomics. *Proc. NatL. Acad. Sci. USA* 107: 8985–8992.

Rindermann, H. 2022. Biological categorization within *Homo sapiens* and its consequences for differences in behavior – or not. *Human Evol.* 37: 139–179.

Risch, N., Burchard, E., Ziv, E., and Tang, H. 2002. Categorization of humans in biomedical research: genes, race and disease. *Genome biol.* 3(7): comment 2007.1–2007.12.

Robert, J. 2025. Pris en tenaille entre wokisme et trumpisme (Caught between Wokism and Trumpism). *Observatoire du Décolonialisme*, 6th February. https://decolonialisme.fr/pris-en-tenaille-entre-wokisme-et-trumpisme/.

Roberts, D. E. 2021. The art of medicine. Abolish race correction. *The Lancet* 397: 17–18.

Rodriguez Mega, E. 2021. The rise of the mixed-race myth in Latin America. *Nature* 600: 374–378.

Rose, S. 2009. Should scientists study race and IQ? NO: Science and society do not benefit. *Nature* 457: 786–788.

Rosenberg, N. A. 2018. Variance-partitioning and classification in human population genetics. pp. 399–403. *In:* Winther, R. G. (ed.). Phylogenetic Inference, Selection Theory, and History of Science: Selected Papers of A. W. F. Edwards with Commentaries. Cambridge: Cambridge University Press.

Rosenberg, N. A. 2020. A Population-Genetic Perspective on the Similarities and Differences among Worldwide Human Populations. *Human Biol.* 92: 135–152.

Rosenberg, N. A., Pritchard, J. K., Weber, J. L., Cann, H. M., Kidd, K. K. et al. 2002. Genetic structure of human populations. *Science* 298: 2381–85.

Rosenberg, N. A., Mahajan, S., Ramachandran, S., Zhao, C., Pritchard J. K. et al. 2005. Clines, Clusters, and the Effect of Study Design on the Inference of Human Population Structure. *Plos Genet.* 1(6): e70. https://doi.org/10.1371/journal.pgen.0010070.

Rosier, F. 2018. Génomique à grande échelle, un bilan en demi-teinte (Large-scale genomics: a mixed assessment) *Le Monde*, 11th September. https://www.lemonde.fr/sciences/article/2018/09/10/genomique-a-grande-echelle-un-bilan-en-demi-teinte_5353150_1650684.html.

Roth T. C., Gallagher C. M., LaDage, L. D., and Pravosudov V. V. 2012. Variation in brain regions associated with fear and learning in contrasting climates. *Brain Behav. Evol.* 79 : 181–190.

Rotimi, C. N. 2004. Are medical and nonmedical uses of large-scale genomic markers conflating genetics and "race"? *Nature Genet. Suppl.* 36 : S43eS47.

Royer, E. 2024. SOS Racisme a 40 ans : grandeur et déclin d'une association jadis dominante ("SOS Racisme is 40 years old: the rise and fall of a once dominant association"). *Valeurs actuelles*, 29th November. https://www.valeursactuelles.com/clubvaleurs/societe/lanniversaire-depasse-de-sos-racisme.

Ruffié, J. 1983. *De la biologie à la culture* (from Biology to Culture). 2nd volume. Flammarion éditeur, Paris, France.

Rühli, F., and Henneberg, M. 2017. Biological Future of Humankind: Ongoing Evolution and the Impact of Recognition of Human Biological Variation. pp. 263–75. *In:* Michel Tibayrenc and Francisco J. Ayala (eds.). *On Human Nature: Biology, Psychology, Ethics, Politics, and Religion.* San Diego: Elsevier/Academic Press.

Rushton, J. P., and Jensen, A. R. 2006. The Totality of Available Evidence Shows the Race IQ Gap Still Remains. *Psychol. Sci.* 17: 921–922.

Sablonnière, B. 2020. ADN, histoire de nos différences (DNA, the story of our differences). Editions Odile Jacob, Paris, France.

Saint Pierre, A., Giemza, J., Alves, I., Karakachoff, M., Gaudin, M. et al. 2020. The genetic history of France. *Eur. J. Hum. Genet.* 28: 853–865.

Saini, A. 2019. Superior: the retour of race science. Beacon Press.

Sarich, V., and Miele, F. 2004. Race: The Reality of Human Differences. Boulder, CO: Westview Press.

Sastre, P. 2021. Phébé – Le paradoxe de l'antiracisme (the Paradox of Antiracism). *Le Point Phébé*, 25th June. https://www.lepoint.fr/phebe/phebe-le-paradoxe-de-l-antiracisme-25-06-2021-2432744_3590.php.

Sauer, N. J. 1992. Forensic anthropology and the concept of race: if races don't exist, why are forensic anthropologists so good at identifying them? *Soc. Sci. Med.* 34: 107–111.

Scerri, E. M. L., Thomas, M. G., Manica, A., Gunz, P., Stock, J. T. et al. 2018. Did our species evolve in subdivided populations across africa, and why does it matter? *Trends Ecol. Evol.* 33: 582–94.

Schaefer, N. K., and Shapiro, B. 2019. New middle chapter in the story of human evolution. *Science* 365: 981–82.

Schwartz, R. S. 2001. Racial profiling in medical research. *N. Engl. J. Med.* 344: 1392–1393.

Sesardic, N. 2013. Confusions about race: A new installment. *Studies in History and Philosophy of Biological and Biomedical Sciences* 44: 287–293.

Serre, D., and Pääbo, S. 2004. Evidence for gradients of human genetic diversity within and among continents. *Genome Res.* 14: 1679–1685.

Shelton. J. F., Shastri, A. J., Ye, C., Weldon, C. H., Filshtein-Sonmez, T. et al. 2021. Trans-ancestry analysis reveals genetic and nongenetic associations with COVID-19 susceptibility and severity. *Nature Genet.* 53: 801–808.

Shi, Y., Niu, Y., Zhang, P., Luo, H., Liu, S. et al. 2023. Characterization of genome-wide STR variation in 6487 human genomes. *Nature Com.* 14: 2092. https://doi.org/10.1038/s41467-023-37690-8.

Shiao, J. L., Bode T., Beyer, A., and Selvig, D. 2012. The genomic challenge to the social construction of race. *Sociolog. Theory* 30: 67–88.

Shukla, P., and Singh, K. K. 2021. Uncovering Mitochondrial Determinants of Racial Disparities in Ovarian Cancer. *Trends in Cancer* 7: 93–97.

Silventoinen, K., Sammalisto, S., Perola, M., Boomsma, D. I., Cornes, B. K. et al. 2003. Heritability of Adult Body Height: A Comparative Study of Twin Cohorts in Eight Countries. *Twin Research* 6: 399–408.

Simonti, C. 2025. Human Genetics Adapting to the Himalayas. *Science* 389: 697.

Smith, L. A., Cahill, J. A., Ji-Hyun Lee, J. H., and Graim, K. 2025. Equitable machine learning counteracts ancestral bias in precision medicine. *Nature Com.* 16: 2144.

Sneath, P. H. A., and Sokal, R. R. 1973. *Numerical Taxonomy. The principle and practice of numerical classification.* (Kennedy, D. and Park R.B., eds.), pp. 537. Freeman, San Francisco.

Sniekers, S., Stringer, S., Watanabe, K., Jansen, P. R., Coleman, J. R. et al. 2017. Genome-wide association meta-analysis of 78,308 individuals identifies new loci and genes influencing human intelligence. *Nature Genet.* 49: 1107–1115.

Southam, L., and Zeggini, E. 2025. Twenty years of genome-wide association studies. *Nature* 641: 47–49.

Sparks, C. S., and Jantz, R. L. 2002. A reassessment of human cranial plasticity: Boas revisited. *Proc. Natl. Acad. Sci. USA* 99: 14636–14639.

Spielman, R. S., Bastone, L. A., Burdick, J. T., Morley, M., Warren, J. et al. 2007. Common genetic variants account for differences in gene expression among ethnic groups. *Nature Genet.* 39: 226–231.

Staeger, E. 2025. Depuis 150 ans, le bassin des femmes rétrécit une évolution qui pourrait bouleverser notre façon d'accoucher (For 150 years, women's pelvises have been shrinking: a development that could revolutionize the way we give birth). *Slate*, 26th June. https://www.slate.fr/sante/bassin-femmes-retrecir-impact-accouchement-cesarienne-evolution?utm_source=firefox-newtab-fr-fr.

Stevens, L. A., Claybon, M. A., Schmid, C. H., Chen, J., Horio, M. et al. 2011. Evaluation of the Chronic Kidney Disease Epidemiology Collaboration equation for estimating the glomerular filtration rate in multiple ethnicities *Kidney Int.* 79: 555–562.

Stoczkowski, W. 2006. L'antiracisme doit-il rompre avec la science ? (Should anti-racism break with science?) *La Recherche* 401: 44–48.

Stoczkowski, W. 2007. Racisme, antiracisme et cosmologie lévi-straussienne. Un essai d'anthropologie réflexive (Racism, Antiracism, and Lévi-Straussian Cosmology. An Essay in Reflexive Anthropology.). *L'Homme*. Revue française d'anthropologie. 182 : April-June. https://journals.openedition.org/lhomme/29406?lang=en#tocto1n2.

Stoneking, M. 2017. The Contribution of Genetic Ancestry From Archaic Humans to Modern Humans. In *On Human Nature: Biology, Psychology, Ethics, Politics, and Religion*, edited by Michel Tibayrenc and Francisco J. Ayala, 55–63. San Diego : Elsevier/Academic Press.

Stoneking, M., and Krause, J. 2011. Learning about human population history from ancient and modern genomes. *Nature Rev. Genet.* 12: 603–614.

Stoppa-Lyonnet, D. 2015. "Race" l'ignorance conduit à la détestation ("Race" : Ignorance leads to hatred). *Le Monde*, 6th October. https://www.lemonde.fr/sciences/article/2015/10/06/race-l-ignorance-conduit-a-la-detestation_4783024_1650684.html.

Strauch-Bonart, L. 2020. Thuram/Bruckner – Existe-t-il un "privilège blanc" ? (Thuram/Bruckner – Is there such a thing as "white privilege"?) *Le Point*, 13th October. https://www.lepoint.fr/postillon/thuram-bruckner-existe-t-il-un-privilege-blanc-13-10-2020-2396139_3961.php.

Stricot, M. 2023. Gènes, culture et évolution : nouvel éclairage sur l'aventure humaine (Genes, culture, and evolution: new insights into the human adventure). *J. CNRS* 9th January. https://lejournal.cnrs.fr/articles/genes-culture-et-evolution-nouvel-eclairage-sur-laventure-humaine.

Stringer, C. B., and Andrews, P. 1988. The origin of modern humans. *Science* 239: 1263–68.

Štrkalj, G. 2007. The Status of the Race Concept in Contemporary Biological Anthropology: A Review. *Anthropologist* 9: 73–78.

Sudmant, P. H., Rausch, T., Gardner, E. J., Handsaker, R. E., Abyzov, A. et al. 2015. An integrated map of structural variation in 2,504 human genomes. *Nature* 526: 75–81.

Sümer, A. P., Rougier, H., Villalba-Mouco, V., Huang, Y., Iasi, L. N. et al. 2024. Earliest modern human genomes constrain timing of Neanderthal admixture. *Nature*, https://doi.org/10.1038/s41586-024-08420-x.

Taguieff, P. A. 1988. La force du préjugé. Essai sur le racisme et ses doubles (The Power of Prejudice: An Essay on Racism and Its Duplicates). Armillaire la découverte Paris, France.

Taguieff, P. A. 2018. "race": un mot de trop ? ("race": one word too many?) CNRS éditions, Paris, France.

Taguieff, P. A. 2020. L'imposture décoloniale. (The decolonial imposture) Editions de l'Observatoire/ Humensis, France.

Taguieff, P. A. 2022. L'antiracisme saisi par l'utopisme. De la mixophilie universelle à la "créolisation du monde" (Anti-racism seized by utopianism. From universal mixophilia to the "creolization of the world"). *Front populaire* 8th December, pp. 52–65.

Taguieff, P. A. 2023. La déconstruction, mot magique et machine de guerre contre la civilisation occidentale (Deconstruction, a magic word and a weapon of war against Western civilization). *Atlantico*, 29th April. https://atlantico.fr/article/decryptage/la-deconstruction-mot-magique-et-machine-de-guerre-contre-la-civilisation-occidentale-culture-universites-wokisme-guerre-culturelle-ideologie-pierre-andre-taguieff.

Taguieff, P. A. 2024. La bêtise se concentre dans les milieux qui collectionnent les "bonnes causes" (Stupidity is concentrated in circles that collect "good causes".). *Tribune juive* 1725: 38: 22nd February.

Taguieff, P. A. 2025. Ni wokisme gauchiste ni antiwokisme trumpiste: l'appel au sursaut de Pierre-André Taguieff (Neither leftist wokism nor Trumpian anti-wokism: Pierre-André Taguieff's call for a resurgence). *L'Express*, 18th March. https://www.lexpress.fr/idees-et-debats/ni-wokisme-gauchiste-ni-antiwokisme-trumpiste-lappel-au-sursaut-de-pierre-andre-taguieff-O3GO5UKIJRHUBI4BRFPQTTL7LA/.

Takezawa, Y. I., Smedley, A., Wade, P., Augustyn, A., Kuiper, K. et al. 2023. Race. *Encyclopaedia Britannica* Updated October 19th. https://www.britannica.com/topic/race-human.

Tate, S. K., and Goldstein, D. B. 2004. Will tomorrow's medicine work for everyone? *Nature Genet. Suppl*. 36: S34–S42.

Templeton, A. R. 2012. *Gene Flow, Haplotype Patterns and Modern Human Origins. In:* eLS. John Wiley & Sons, Ltd: Chichester. DOI: 10.1002/9780470015902.a0020795.pub2.

Templeton A. R. 2013. Biological races in humans. *Stud. Hist. Philos. Biol. Biomed. Sci.*

Templeton, A. R. 2017. World Dispersals and Genetic Diversity of Mankind: The Out-of-Africa Theory and Its Challenges. *In: On Human Nature: Biology, Psychology, Ethics, Politics, and Religion* edited by Michel Tibayrenc and Francisco J. Ayala, 65–83. San Diego : Elsevier/Academic Press.

Tennessen, J. A., Bigham, A. W., O'Connor, T. D., Fu, W., Kenny, E. E. et al. 2012. Evolution and Functional Impact of Rare Coding Variation from Deep Sequencing of Human Exomes. *Science* 337: 64–69.

The 1000 Genomes Project Consortium 2015. A global reference for human genetic variation. *Nature* 526: 68–74.

The international journal of science. 2023. Why *Nature* is updating its advice to authors on reporting race or ethnicity. *Nature* 616: 219.

The UK10K Consortium. 2015. The UK10K project identifies rare variants in health and disease. *Nature* 526: 82–89.

The UK Biobank Whole-Genome Sequencing Consortium 2025. 692 *Nature* 645: 692–709.

Thomas, C. E., and Peters, U. 2025. Genomic landscape of cancer in racially and ethnically diverse populations. *Nature Rev. Genet*. 26: 336–349.

Tibayrenc, M. 1979. Les isoenzymes et l'Entomologie médicale (isoenzymes and medical entomology). *Cah. ORSTOM sér. Ent. méd. Parasitol*., 17 (4): 249–256.

Tibayrenc, M. 2009. Multilocus enzyme electrophoresis for parasites and other pathogens. *In*: Caugant, D. A. (ed.). Methods in Molecular Biology: Molecular Epidemiology of Microorganisms. Humana Press.

Tibayrenc, M. 2017a. Human Population Variability and Its Adaptive Significance. pp. 13–25. *In:* Michel Tibayrenc and Francisco J. Ayala (eds.). *On Human Nature: Biology, Psychology, Ethics, Politics, and Religion.* 85–109. San Diego : Elsevier/Academic Press.

Tibayrenc, M. 2017b. The Race/Ethnic Debate: An Outsider's View. In *On Human Nature: Biology, Psychology, Ethics, Politics, and Religion*, edited by Michel.

Tibayrenc and Francisco J. Ayala, 633–49. San Diego : Elsevier/Academic Press.

Tibayrenc, M. 2017c. Human Intergroup Variation and Disease Genetics. In *On Human Nature: Biology, Psychology, Ethics, Politics, and Religion*, edited by Michel Tibayrenc and Francisco J. Ayala, 161–75. San Diego : Elsevier/Academic Press.

Tibayrenc, M. 2019. Response to ASHG: Science and politics should be mutually sanctuarized. *Am. J. Human Genet.* 104: 774–75.

Tibayrenc, M. 2022. La diversité de notre espèce ne veut pas dire hiérarchisation. (The diversity of our species does not imply hierarchy). Le *Droit de Vivre*, 28th June. https://www.leddv.fr/entretien/michel-tibayrenc-la-diversite-de-notre-espece-ne-veut-pas-dire-hierarchisation-20220628?fbclid=IwAR0Uj71nN-zg4W8lSuq3SNI2bYzJGJn4JVKbQo7nATNFXGeST_s_QNQgFls.

Tibayrenc, M., and Ayala, F. J. 2013. How clonal are *Trypanosoma* and *Leishmania*? *Trends Parasitol.* 29: 264–69

Tibayrenc, M., and Ayala, F. J. 2017a. Relevant units of analysis for applied and basic research dealing with neglected transmissible diseases: The predominant clonal evolution model of pathogenic microorganisms. *Plos Neglect. Trop. Dis.* 11(4): e0005293. https://doi.org/10.1371/journal.pntd.0005293.

Tibayrenc, M., and Ayala, F. J. 2017b. On Human Nature: Biology, Psychology, Ethics, Politics, and Religion. San Diego : Elsevier/Academic Press.

Tibayrenc, M., and Ayala, F. J. 2020. What makes us humans. Nova Science publishing New York.

Tishkoff, S. A., and Kidd, K. K. 2004. Implications of biogeography of human populations for "race" and medicine. *Nature Genet. Suppl.* 36: S21–S27.

Tournier, V., and Robert, J. 2025. Stand up for science ! Oui, mais pour tous les scientifiques victimes d'ostracisme (Stand up for science ! Yes, but for all scientists who are victims of ostracism...) *Observatoire du décolonialisme* 11th March. https://decolonialisme.fr/stand-up-for-science-oui-mais-pour-tous-les-scientifiques-victimes-dostracisme/.

Tung, J. 2024. Understanding human uniqueness in the pre-genomic era. *Nature Rev. Genet.* 25: 459.

UNESCO statement on race. 2022. Encyclopedia of Race and Racism. Encyclopedia.com. 10th June <https://www.encyclopedia.com>.

Vallois, H. V. 1976. *Les races humaines* (Human Races). Paris, France : Presses Universitaires de France.

Van Laere, J. E. 1961. *Éléments de psychiatrie* (Elements of Psychiatry). Paris, France : Éditions Médicales Flammarion.

Van Valen, L. 1973. A new evolutionary law. *Evolut. Theory* 1: 1–30.

Vasseur, E., and Quintana-Murci. L. 2013. The impact of natural selection on health and disease: uses of the population genetics approach in humans. *Evolutionary Applications* 6: 596–607.

Verma, A., Huffman, J. E., Rodriguez, A., Conery, M., Liu, M. et al. 2024. Diversity and scale: Genetic architecture of 2068 traits in the VA Million Veteran Program. 2024. *Science* 385: eadj1182.

Vey, T. 2018. Les pensées racistes d'Einstein dévoilées dans ses carnets de voyages. (Einstein's racist thoughts revealed in his travel journals) *Le Figaro*, 14th June. http://premium.lefigaro.fr/sciences/2018/06/14/01008-20180614ARTFIG00306-les-pensees-racistes-d-einstein-devoilees-dans-ses-carnets-de-voyages.php.

Viguier, A. 2023. Ré-écrire l'histoire : comment le pouvoir en Inde cherche à diffuser une vision exclusivement hindoue de son passé (Rewriting history: how power in India seeks to spread an exclusively Hindu vision of its past). *The Conversation*, 12th July. https://theconversation.com/re-ecrire-lhistoire-comment-le-pouvoir-en-inde-cherche-a-diffuser-une-vision-exclusivement-hindoue-de-son-passe-208784.

Villard, A., Grauby-Vermeil, G., and Vidard, M. 2015. La réponse d'Axel Kahn et Pierre-Henri Gouyon à Nadine Morano (Axel Kahn and Pierre-Henri Gouyon's response to Nadine Morano). France Inter, 30th September. https://www.franceinter.fr/idees/la-reponse-d-axel-kahn-et-pierre-henri-gouyon-a-nadine-morano.

Vincent, C. 2013. Fin de "race" (End of Race). *Le Monde*, 27th June. https://www.lemonde.fr/culture/article/2013/06/27/fin-de-race_3437989_3246.html.

vonHoldt, B. M., Pollinger, J. P., Lohmueller, K. E., Han, E., Parker, H. G. et al. 2010. Genome-wide SNP and haplotype analyses reveal a rich history underlying dog domestication. *Nature* 464: 898–902.

Wade, L. 2018. Ancient DNA untangles South Asian roots. *Science* 360: 252.

Wade, L. 2021a. Anthropologists grapple with identifying remains by race. *Science* 374: 386–387.

Wade, L. 2021b. The ghosts in the museum. *Science* 373: 149–151.

Wade, N. 2014. A troublesome inheritance. Genes, race and human history. Penguin books, New York.

Wang, Y., He, Y., Shi, Y., Qian, D. C., Gray, K. J. et al. 2024. Aspiring toward equitable benefits from genomic advances to individuals of ancestrally diverse backgrounds. *Am. J. Human Genet.* 111: 809–824.

Weidenreich, F. 1946. *Apes, Giants, and Man*. Chicago : University of Chicago Press.

Westby, A., Okah, E., and Ricco, J. 2020. Race-Based Treatment Decisions Perpetuate Structural Racism. *Am. Family Phys.* 102: 136–137.

Wilde, S., Timpson, A., Kirsanow, K., Kaiser, E., Kayser, M. et al. 2014. Direct evidence for positive selection of skin, hair, and eye pigmentation in Europeans during the last 5,000 y. *Proc. Natl. Acad. Sci. USA* 111: 4832–4837.

Williams, D. R., González, H. M., Neighbors, H., Nesse, R., Abelson, J. M. et al. 2007. Prevalence and Distribution of Major Depressive Disorder in African Americans, Caribbean Blacks, and Non-Hispanic Whites. Results From the National Survey of American Life. *Arch. Gen. Psychiatry* 64: 305–315.

Wilson, E. O. 1978. *On Human Nature*. Harvard University Press, Cambridge, Massachssets.

Winegard, B., Winegard, B., and Anomaly, J. 2020. Dodging Darwin: Race, evolution, and the hereditarian hypothesis. *Pers. Individ. Differ.* https://doi.org/10.1016/j.paid.2020.109915.

Winther, R. G. 2018. A Beginner's Guide to the New Population Genomics of *Homo sapiens*: Origins, Race, and Medicine. *Harv. Rev. Philos*. XXV: 1–18.

Wojcik, G. L. 2025. Eugenics is on the rise again: geneticists must take a stand. *Nature* 641: 37–38.

Wolpoff, M. H. 2009: How Neandertals Inform Human Variation. *Am. J. Phys. Anthropol.* 139: 91–102.

Wood, B., and Boyle, E. 2017. Hominins: context, origins, and taxi diversity. pp. 17–44. *In*: Michel Tibayrenc and Francisco J. Ayala (eds.). *On Human Nature: Biology, Psychology, Ethics, Politics, and Religion*. San Diego : Elsevier/Academic Press.

Wright, J. L., Davis, W. S., Joseph, M. M., Ellison, A. M., Heard-Garris, N. J. et al. 2022. Eliminating Race-Based Medicine *Pediatrics* 150, number 1, July 2022: e2022057998.

Wright, S. 1951. The genetical structure of populations. *Annals of Eugenics* 15: 323–54.

Wynshaw-Boris, A. 2020. American Society of Human Genetics Statement Regarding Concepts of "Good Genes" and Human Genetics. *ASHG News*, 24th September. https://www.ashg.org/publications-news/ashg-news/statement-regarding-good-genes-human-genetics/#:~:text=Genetics%20demonstrates%20that%20humans%20cannot,ancestry%20have%20no%20scientific%20evidence.

Xie, N. 2024. Building a catalogue of short tandem repeats in diverse populations. *Nature Rev. Genet.* 25: 457.

Yong, E. 2013. Chinese project probes the genetics of genius. *Nature* 497: 297–299.

Yudell, M., Roberts, R., DeSalle, R., and Tishkoff, S. 2016. Taking race out of human genetics. *Science* 351: 564–65.

Yudell, M., Roberts, D., DeSalle, R., Tishkoff, S., and 70 signatories. 2020. NIH must confront the use of race in science. *Science* 369: 1313–1314.

Zampieri, F. 2017. The impact of modern medicine on human evolution. pp. 707–27. *In*: Michel Tibayrenc, and Francisco J. Ayala (eds.). *On Human Nature: Biology, Psychology, Ethics, Politics, and Religion*. San Diego: Elsevier/Academic Press.

Zietsch, B. P., de Candia T. R., and Keller, M. C. 2015. Evolutionary behavioral genetics. *Curr. Opinion in Behavioral Sci.* 2: 73–80.

Zou, J., Gichoya, J. W., Ho, D. E., and Obermeyer, Z. 2023. Implications of predicting race variables from medical images. *Science* 381: 149–150.

Index

Author's Biography

Michel Tibayrenc, MD, PhD, is a Director of Research Emeritus at the French Institut de Recherche pour le Développement, with 45 years of expertise in the genetics and evolution of infectious diseases. He is the founder and editor-in-chief emeritus of the journal Infection, Genetics and Evolution, the editor-in-chief of the new journal Diseases: Biology, Genetics and Socioecology, and established the MEEGID international congresses, which he organized from 1996 to 2021.

With an H-index of 50, Dr. Tibayrenc has authored over 200 international papers and published 7 major scientific books. His notable works include Genetics and Evolution of Infectious Diseases, several volumes on human nature co-edited with Francisco J. Ayala, and Race and Racism: Biology, Ethics, Politics (CRC Press).

He is the co-founder and scientific adviser of the Bolivian Society Genetics. His career remains defined by a profound contribution to the evolutionary study of transmissible diseases.

For Product Safety Concerns and Information please contact our EU representative GPSR@taylorandfrancis.com Taylor & Francis Verlag GmbH, Kaufingerstraße 24, 80331 München, Germany